WITHDRAWN
UTSA Libraries

THE CIVIL WAR:

A SOLDIER'S VIEW

THE CIVIL WAR:

A SOLDIER'S VIEW

A COLLECTION OF CIVIL WAR WRITINGS

BY COL. G. F. R. HENDERSON

Edited by Jay Luvaas

THE UNIVERSITY OF CHICAGO PRESS
Chicago and London

Library of Congress Catalog Card Number: 57-11209

THE UNIVERSITY OF CHICAGO PRESS, CHICAGO 60637
THE UNIVERSITY OF CHICAGO PRESS, LTD., LONDON W. C. 1

©*1958 by The University of Chicago. All rights reserved
Published 1958. Second Impression 1968. Printed in the United
States of America*

PREFACE

G. F. R. Henderson is no stranger to those interested in the American Civil War. For half a century his monumental biography of Stonewall Jackson has been acclaimed as the best study of the famous Confederate general and a classic in military literature.

Yet Henderson's other writings on the Civil War are comparatively unknown. *The Campaign of Fredericksburg* (chapter ii) is a scarce work, and the interpretative essays on the men and events of 1861–65 (chapters iv–vii), contained in *The Science of War*, a posthumous collection of Henderson's articles and speeches, have never been given the attention they deserve. "Stonewall Jackson's Place in History" (chapter viii) was written before Henderson had made his reputation in America and lies buried in Mrs. Jackson's forgotten biography of her husband.

These writings deserve a better fate. Not only are they worthwhile reading in themselves, for Henderson was a thoughtful and well-informed student of the Civil War; they also enable us to trace Henderson's footsteps in preparing himself to write *Stonewall Jackson and the American Civil War*, and they demonstrate, more clearly than the biography itself, the fact that Henderson's interest in the American campaigns was by no means purely academic. He exercised considerable influence in the British army, and it was through his Civil War studies that Henderson presented many of his own ideas on vital military questions of the day. In chapters i, iii, and ix, I have endeavored to explain Henderson's interest in the Civil War, to determine what military lessons he hoped to learn from the American conflict, and to suggest the nature and extent of his influence in Britain.

Save for chapter ii, the original text has required but little editing. Henderson's account of Fredericksburg was written before most of the official records of that battle had been made available; therefore his pages are not free from minor errors of fact. Detected errors have been called to the attention of the reader, misspellings of names and places have been corrected, and occasionally it has seemed desirable to supplement Henderson's interpretations with the views of more recent authorities.

Obscure military terms and expressions have been explained whenever it was felt necessary. Military writers and issues have been identified only when they pertain to Henderson's career or to the Civil War. No attempt has been made to identify all the personalities and historical illustrations mentioned in the text.

Henderson's footnotes in chapter iv have been changed to conform to modern usage, and one footnote in chapter vi originally belonged in the text. In rare instances Henderson's punctuation has been altered for the sake of consistency.

The Campaign of Fredericksburg was first published in 1886, at Henderson's own expense. At least two subsequent editions were published in England by Gale and Polden, Limited, well-known publishers of military works. I gratefully acknowledge permission of this firm to reproduce *The Campaign of Fredericksburg* in the present volume.

Similarly, Longmans, Green and Company, publishers of *The Science of War,* have kindly granted permission for the inclusion of chapters iv through vii.

Substantial portions of chapters i, iii, and ix appeared originally in *Military Affairs* and are used here with the consent of Captain Victor Gondos, Jr., editor.

Many friends have given advice and encouragement. Among those to whom I am deeply indebted are Wilbor Kraft, who first suggested a new edition of Henderson's Civil War writings; Stuart E. Brown, Jr., who called my attention to the original Henderson sketches in the Handley Library; and Dr. David C. Mearns, chief of the Manuscripts Division of the Library of Congress, who made available microfilm copies of Henderson's correspondence with Major Jed Hotchkiss.

Two in particular deserve special thanks. Captain B. H. Liddell Hart, always a helpful and generous friend, offered sound advice and proved to be as important a source of information to me regarding the British army as Jed Hotchkiss was to Henderson in supplying information on the life and campaigns of Stonewall Jackson. Above all, I wish to acknowledge my debt to a stimulating teacher and a valued friend, Professor Theodore Ropp, of Duke University, who first introduced me to Henderson, guided me in research, and made many valuable criticisms of the introductory chapters.

Finally, endless thanks are due my wife, tireless proofreader and uncomplaining Civil War widow.

CONTENTS

List of Illustrations   xi

I. Introduction   1

II. "The Campaign of Fredericksburg"   9

III. Henderson and the American Civil War   120

IV. "Battles and Leaders of the Civil War"   130

V. "The American Civil War, 1861–1865"   174

VI. "The Battle of Gettysburg"   225

VII. "The Campaign in the Wilderness of Virginia, 1864"   254

VIII. "Stonewall Jackson's Place in History"   284

IX. The Henderson Legacy   301

Index   315

LIST OF ILLUSTRATIONS

MAPS   PAGE

I. N.E. Virginia, Theatre of Campaign   15

II. Country round Fredericksburg, Showing Confederate Dispositions along the Rappahannock, December 3–12, 1862   23

III. Right Bank of Rappahannock, with Federal Corps, Night of December 12, 1862   45

IV. Approximate Positions of Troops at 9.15 a.m., December 13, 1862   60

V. Vicinity of Brandy Station, Showing Battlefield of June 9, 1863   215

VI. The Campaign of Gettysburg   239

VII. The Campaign in the Wilderness   276–77

SKETCHES

I. Infantry Shelter Trench and Stone Wall Prepared for Defence at Fredericksburg   49

II. Approximate Positions and Strength of the Different Divisions at 4 p.m., December 13, 1862   87

FACING PAGE

Map of Virginia and Maryland   226

INTRODUCTION I

In 1886 there appeared in England a small volume entitled *The Campaign of Fredericksburg*, written by an anonymous "Line Officer." To the average reader this book was probably just another campaign history, better written than most tactical studies but otherwise no different from many similar volumes in the regimental library. But to those concerned with the education of the British officer this book had a unique appeal. Written with intelligence and unusual insight and filled with thoughtful observations on the military significance of the campaign, it represented a skilful blending of personal knowledge of the terrain, careful study of the available sources, and a lively and readable style. Colonel (later Sir Frederick) Maurice, professor of military art and history at the Staff College, reviewed the book enthusiastically, and, recalling that Sir Garnet Wolseley, at that time the adjutant general, had once visited the Confederate army, he forwarded his copy to Wolseley. Wolseley read it, liked it, and made inquiries about the identity of the author. He was charged with the military education of British officers and had use for an able instructor.[1]

1. Lt. Col. F. Maurice, *Sir Frederick Maurice: A Record of His Work and Opinions* (London, 1913), p. 64; Major General Sir F. Maurice and Sir George Arthur, *The Life of Lord Wolseley* (New York, 1924), p. 236.

"Line Officer" proved to be Captain G. F. R. Henderson, York and Lancaster Regiment. Henderson, then thirty-two years of age, had attended Oxford University on a scholarship in history before entering the army in 1877. In the Egyptian campaign of 1882, he had distinguished himself by rushing an enemy redoubt at Tel-el-Kebir. Assigned to garrison duty in Bermuda the following year, Henderson had subsequently been transferred to Halifax, where he had spent his leave tramping over many of the battlefields in Virginia and Maryland. Upon his return to England, Henderson had applied for and received assignment to the Ordnance Department, and this assignment provided him with sufficient time and income to continue his study of the Civil War.[2]

Wolseley was so impressed with Henderson that he appointed him to the faculty of the Cadet School at Sandhurst. An elaborate tactical study of the battle of Spicheren, published in 1891,[3] led to another advancement, and in the following year Henderson was appointed Maurice's successor at the Staff College. Rarely has publication led to such rapid promotion, even in an academic atmosphere.

The fact that *The Campaign of Fredericksburg* gave Henderson his opportunity only partially explains the significance of this book. Henderson was actually the first English officer to undertake a serious study of the Civil War after 1870, when the Prussian army had stunned the military world by overrunning France in a series of swift, decisive battles. This defeat of what was then regarded as the foremost military power in Europe had caused all eyes to focus on the German army. A new military literature had appeared, stressing questions of tactical interest and for the most part disregarding campaigns conducted before 1866 on the broad assumption that "you can't draw any deduc-

---

2. Information on Henderson's career is found in E. M. Lloyd, "Henderson," *Dictionary of National Biography*, Suppl. II, 240–41; Lt. Col. R. M. Holden, "Lt. Col. G. F. R. Henderson, C.B., In Memoriam," *Journal of the Royal United Service Institution*, XLVII (April, 1903), 375–82, and an obituary, probably by the same author, in the *Times*, March 7, 1903. There is also a biographical appreciation by Field Marshal Earl Roberts in Henderson, *The Science of War*, ed. Captain Neill Malcolm (London, 1906), pp. xiii–xxxviii (cited hereafter as Roberts, "Memoir"). Many letters from Henderson to Major Jedediah Hotchkiss are included in the Jed Hotchkiss Papers, Library of Congress.

3. G. F. R. Henderson, *The Battle of Spicheren, August 6th, 1870, and the Events That Preceded It: A Study in Practical Tactics and War Training* (Aldershot, 1891).

tions from what occurred before the introduction of the breech-loader."[4] Since 1870, students at the Staff College had been fed a regular diet of Moltke and his campaigns. In the 1880's the instructor in military history was Lieutenant Colonel Lonsdale Hale, a pedant who had, in the words of General J. F. C. Fuller, "so minutely studied" the Franco-Prussian War that "at any moment, he could inform an inquirer of the exact position of all the German and French units down to companies at any given time in any battle."[5] This may be an extreme example, but it does serve to illustrate the importance most British officers attached to the campaigns of 1870–71. The Civil War was treated only casually in the textbooks on tactics[6] and was practically ignored in the military journals. So convincing had been the Prussian triumph that English soldiers ceased to study the Civil War for nearly a generation.

There had been a time when the Civil War was a popular subject at the Staff College. The first commandant, Colonel (later General Sir Patrick) MacDougall, had mentioned some of the lessons of the Civil War in his *Modern Warfare as Influenced by Modern Artillery* (1864).[7] Captain Charles Cornwallis Chesney, professor of military history at the Staff College in the 1860's, had used the American campaigns to illustrate his lectures; indeed, Chesney first made his reputation as a military critic through his writings on the Civil War.[8] And Sir Edward Hamley, Chesney's predecessor at the Staff College, likewise had ap-

4. Lt. F. Maurice, *The System of Field Manoeuvres* (London, Wellington Prize Essay for 1872), p. 10; *War* (London, 1891), p. 110.

5. Major General J. F. C. Fuller, *The Army in My Time* (London, 1935), p. 123. According to Sir John Adaye, who attended the Staff College in 1885–86, Lonsdale Hale "had a 'cramming' establishment in Kensington, and . . . was, I suppose, the greatest authority in England at that time upon the Franco-German War, the lessons of which were still held to be of the greatest importance for all soldiers" (*Soldiers and Others I Have Known* [London, 1925], pp. 136–37).

6. See Lt. Col. Clery, *Minor Tactics* (5th ed.; London, 1880), *passim;* and Colonel Robert Home, *A Précis of Modern Tactics,* revised and rewritten by Lt. Col. Sisson C. Pratt (London, 1892), *passim.*

7. (London), pp. 5, 14–15, 23, 135, 272, 351, 424.

8. Chesney was probably best known to English soldiers for his *Waterloo Lectures* (London, 1874). See Stanley Lane-Poole, "Charles Cornwallis Chesney," *Dictionary of National Biography,* IV, 195; Whitworth Porter, *History of the Corps of Royal Engineers* (London, 1889), II, 499. Chesney's most important work on the Civil War is entitled *Campaigns in Virginia, Maryland, . . .* (2 vols.; London, 1864–65), but he also wrote biographical sketches about several of the American generals and he lectured on Sherman's Atlanta campaign. See p. 126 n., below.

preciated the military significance of the Civil War. Hamley, "for generations the strategic pedagogue of the British Army," had contributed several articles about the war to *Blackwood's Magazine*, and in his monumental *Operations of War* (1866) he gave appropriate attention to the Civil War campaigns.[9]

But the influence of these men, at least in so far as the American Civil War was concerned, was negligible after 1871 as British officers carefully scrutinized the German campaigns and made a methodical—at times almost a dogmatic—study of German military theories and institutions. Even the Cardwell Reforms, brought about because of military weaknesses revealed in the Crimean War (1854–56) and the Indian Mutiny (1857) and stimulated by the influence of men like Chesney and Hamley, were not altogether free from the German influence. Although essentially English in origin and designed to correct evils peculiar to the British army, these reforms incorporated some ideas from the Germans, particularly some features of the newly adopted short service and enlistment on a territorial basis. Army maneuvers were instituted in 1871; and even the Prussian *Pickelhaube* replaced the French *képi* in most British regiments, a "new look" reflecting the change in military thinking.[10]

Inevitably a reaction set in against this exaggerated study of everything German. By the mid-1880's there were many who wondered if other wars might not offer more valuable lessons to a colonial power. Germany, these new voices argued, was a land power dependent upon a conscript army for protection, while Britain's first line of defense was still the Royal Navy, her small volunteer army having the double responsibility for home defense and policing the empire. Maurice and Wolseley were among those who now looked beyond the period of the German wars, pointing to lessons "which have in no wise been dimin-

9. Cyril Falls, *A Hundred Years of War* (London, 1953), p. 29; General Sir Edward Bruce Hamley, *The Operations of War Explained and Illustrated* (London, 1866), *passim;* Alexander Innes Shand, *The Life of General Sir Edward Bruce Hamley* (Edinburgh and London, 1895), I, 135–91; Mrs. Gerald Porter, *John Blackwood* (Vol. III of *William Blackwood and His Sons* [London, 1898]), p. 268. Hamley wrote these articles anonymously, and only one, "Books on the American War," *Blackwood's Edinburgh Magazine*, XCIV (December, 1863), has been identified as coming from his pen.

10. Sir George Arthur, *From Wellington to Wavell* (London, [1942]), p. 72; General Sir Robert Biddulph, *Lord Cardwell at the War Office* (London, 1904), pp. 189, 212–13; James Laver, *British Military Uniforms* (London, 1948), p. 21.

ished in value by the changes which have come over the face of war." Maurice had a high estimate of the lessons "to be derived by a careful student from the American war." Lord Wolseley, who always wrote with the problems of England uppermost in his mind, strongly advised his officers to "copy the Germans as regards work and leave their clothes and their methods alone."[11] From far-off India came an anonymous appeal to look again at the Civil War, in many respects "a far more likely source of information . . . than the histories of struggles between French and Germans." "A True Reformer" complained that "for 15 years the tactics of timidity have been dinned into our ears, and month after month our officers have been condemned to pass examinations in books compiled from others which have long since found a resting place in the dusty cellars of German publishers."[12]

It was to remedy this situation that Henderson wrote *The Campaign of Fredericksburg*, a book designed to serve as a tactical text for officers of the English Volunteers. This organization, a sickly child of the invasion scare in 1859, was assigned an important role in existing plans for home defense. Unfortunately the Volunteer regiments varied greatly in training and quality, and very little machinery existed whereby they might be swiftly mobilized in time of war.[13] Aware that if the Volunteers were ever called into service they might have to face a Continental army superior both in manpower and in training, Henderson wanted to improve the practical and theoretical training of the Volunteer officers. He believed that a sound knowledge of military history was the best substitute for actual battle experience, and he was satisfied that if Volunteer officers would study past campaigns with intelligence, they would find themselves "instinctively doing the right thing."[14]

11. Maurice, *War*, pp. 12, 107; Maurice and Arthur, *Wolseley*, p. 222. See also Colonel F. Maurice, "How Far the Lessons of the Franco-German War Are Out of Date," *United Service Magazine*, X, N.S. (March, 1895), 555–77; Miller T. Maguire, "Our Art of War as Made in Germany," *ibid.*, XIII (May–June, 1898), 124–34, 280–91; Spenser Wilkinson, "Introduction," W. Birkbeck Wood and Major J. E. Edmonds, *A History of the Civil War in the United States, 1861–65* (New York, 1905), p. xiv.

12. "A True Reformer," *Letters on Tactics and Organization* (Calcutta, 1888), pp. 299, 312. "A True Reformer" was Captain F. N. Maude, a well-known English military writer at the turn of the century.

13. Colonel John K. Dunlop, *The Development of the British Army, 1899–1914* (London, 1938), pp. 55–66.

14. See p. 105, below.

Unlike the majority of those who had studied the Prussian campaigns in elaborate detail, Henderson was not looking for any new or specific military information. While he frequently mentioned the "lessons" to be derived from a study of the American conflict, he used the word to mean "instruction." Napoleon himself had stated that the only way to learn the science of war was to read and reread the campaigns of the Great Captains. Henderson likewise preferred this system to "the tabulated maxims and isolated instances of the text-books," and he considered the battle of Fredericksburg especially suitable for his purposes because both armies had been composed largely of unprofessional soldiers similar to the Volunteers. Surely the experiences of the Americans in 1862 would give the Volunteer officers some insight into the problems they were likely to face.[15]

The "lessons" which this thoughtful English soldier tried to communicate in his account of Fredericksburg all center in his conception of the role of the officer in modern battle. European military observers who had witnessed the Civil War armies in action had noticed that the battles had tended to escape the effective control of the commanding general.[16] Henderson also recognized this apparent trend, which left greater tactical independence in battle to the subordinate, for he emphasized the point that, at Fredericksburg, Lee, having brought his troops to the selected positions, first notified his lieutenants of his general plan "and then gave frankly into their hands the conduct of the fight." Henderson concluded:

> The lesson should be impressed upon all officers who may have the direction of any military operation, however insignificant. Make up your mind clearly as to the course you intend to pursue. Let your plan be as simple as possible; ... take care that it is prosecuted with the utmost determination, and, if your subordinates are qualified to command, leave them to themselves, and beware of unnecessary interference. Be determined at the same time ... to enforce prompt co-operation towards the end in view.[17]

15. See pp. 9, 12, below.
16. Lt. Col. James Arthur Lyon Fremantle, *Three Months in the Southern States* (New York, 1864), p. 260; Major Justus Scheibert, *Der Bürgerkrieg in den Nordamerikanischen Staaten: Militairisch beleuchtet für den deutschen Offizier* (Berlin, 1874), p. 39. Both Fremantle, an officer in Her Majesty's Coldstream Guards, and Scheibert, the official military observer from Prussia, accompanied Lee's headquarters staff during the Gettysburg campaign.
17. See p. 68, below.

Henderson noted the use of intrenchments at Fredericksburg, which furnished "another proof that good infantry, sufficiently covered . . . is, if unshaken by artillery and attacked in front alone, absolutely invincible." He recommended a well-directed fire by volleys as the best defense against frontal attacks, and he considered a movement against the enemy flank or rear as the most likely to succeed against fortified positions. These views harmonized with official British doctrine, for in the years immediately preceding the South African War (1899–1902) volley-firing was the "backbone of all musketry training."[18]

There is some evidence to indicate that Henderson, at this date, underrated the value of modern firepower. He thought Lee's comments on the use of the breech-loader "worth consideration." "What we want," he quoted Lee as saying, "is a firearm that cannot be loaded without a certain loss of time; so that a man learns to appreciate the importance of his fire."[19] He attributed the use of intrenchments to "the wooded nature of the country," and explained that cavalry so often found it necessary to fight on foot with the rifle because the terrain afforded "few facilities for cavalry manoeuvres on a large scale."[20]

According to Henderson, the object lessons of the battle of Fredericksburg, at least in so far as the English Volunteers were concerned, could be summarized in two words—leadership and discipline. The conditions of modern battle made it necessary for junior officers to assume more responsibility; flank marches and volley-firing could be successfully executed only through strict discipline. Henderson admired "the splendid fighting qualities" of the American soldiers. He saw no reason why the Volunteers *"if knit together by strict discipline and led by well-trained officers,"* could not "excel even Lee's battalions in mobility and efficiency."[21]

Henderson's absorbing account of the battle probably will cause many readers to overlook the didactic purpose of *The Campaign of Fredericksburg* and to judge the book on its historical merits alone. The student of the Civil War may find minor

18. See pp. 76, 103–4, 108, below. See also Fuller, *The Army in My Time*, pp. 69–70; Dunlop, *The Development of the British Army*, p. 37.
19. See p. 108, below.
20. See pp. 35–36, below.
21. See p. 119, below.

discrepancies between Henderson's version and later studies: casualty figures are often inflated, and the text is not entirely free from factual errors. It should be remarked that Henderson wrote before publication of the pertinent volume of the *Official Records*[22] and that much of his account was borrowed openly from Chesney's *Campaigns in Virginia*.

But although Henderson himself regarded his book primarily as an amplification of the chapter on Fredericksburg contained in Chesney's earlier work, it was, in fact, much more than that. *The Campaign of Fredericksburg* represents a thorough synthesis of the facts enriched by a perceptive soldier's wide knowledge of military history and his familiarity with the battlefield. After nearly a century, what does it matter if Henderson antedated by one day Lee's dispatch summoning Stonewall Jackson from the Shenandoah Valley or that he claimed Hancock's Union division lost "156 officers and 2,013 men" before Marye's Heights rather than the official casualty returns of 2,013 *including* 156 officers? What is important is the fact that *The Campaign of Fredericksburg* was written at all. Not only is it the book that brought Henderson his opportunity; it is also, according to the foremost authority on Lee and the Army of Northern Virginia, "the best study" of the battle, even today.[23] It is Henderson's style that gives life to the battle, and his authoritative interpretation that makes it intelligible.

22. U.S. War Department, *The War of the Rebellion: A Compilation of the Official Records of the Union and Confederate Armies*, 70 vols. in 128 (Washington, 1880–91). Ser. 1, Vol. XXI, containing reports and correspondence relating to the Fredericksburg campaign, was published in 1888.
23. Douglas Southall Freeman, *Lee's Lieutenants: A Study in Command* (New York, 1944), III, 822. Henderson himself had a more modest opinion of the merits of his first book. He confessed to his favorite American correspondent, "I have no doubt that you will find it full of inaccuracy in details, and, in any case, it is scarcely worth your reading, tho [it is] the only way I have, however, of expressing my hearty thanks for your kindness and courtesy" (Henderson to Jed Hotchkiss, April 24, 1892, Hotchkiss Papers).

THE CAMPAIGN OF FREDERICKSBURG

AUTHOR'S PREFACE

The intelligent study of the actual occurrences of warfare is, next to experience in the field, the surest means of attaining a knowledge of the theory and principles of the military art. Half a dozen great battles, studied in detail and impressed upon the understanding, will give the mind a firmer grasp of tactical principles than the tabulated maxims and isolated instances of the text-books. Not that text-books are to be neglected; from them the elementary rules must be learnt, but only from the study of actual operations and the working-out of practical problems on the map and drill-ground is that instinctive recognition of when and where to apply those rules to be acquired, lacking which no officer can be thoroughly efficient and reliable. To aid the attainment of this desideratum by officers of volunteers, the following pages have been compiled.

Whether the State, as regards the training of such officers, does all in its power, need not be discussed here. However that may be, in any scheme of education much must be left to the individual; and as far as soldiering is concerned, much may be compassed independently of the ordinary sources of instruction.

The numbers who yearly pass the prescribed examination testify to the increasing interest taken in tactical study; but the necessity of such study cannot be too often asserted, for it is very doubtful if the majority of officers—even of the regular forces—as yet appreciate the paramount importance of the knowledge and habit of mind which can be acquired by this means.

Englishmen are too prone to over-estimate the value of physical strength and natural soldierly qualities, too ready to content themselves with looking back at past glories, to believe that the grand endurance of Waterloo, the terrible valour of Inkerman, the heritage of our race, will always of themselves ensure us the victory. But in the face of the high training of foreign armies, it is not enough to rely on national characteristics alone, as the war of 1870–71 abundantly proves. Without altogether concurring in the lately expressed dictum of a foreign critic, that "personal bravery is, at the present time, much less needed than military understanding," it must still be admitted that as military education has kept pace on the Continent with the advancing culture of the century, it behoves English soldiers to take care that they do not fall short in this respect. If ever our volunteers are called upon to take the field, they will find themselves, possibly at a very few days' notice, face to face with a picked army of professional soldiers, led by men who have given their whole time to the study of military science. Moreover, the new tactics necessitated by the power of modern fire-arms have increased the responsibilities of each individual officer. Nevertheless, there is no need to take alarm at the prospect. For regimental officers the sphere of necessary study is but limited; and, though the subjects must be thoroughly mastered and thought out, they are well within the scope of ordinary intelligence and application.

I make no apology for inserting here two extracts from General Sir James Shaw Kennedy's "Notes on Waterloo." Speaking of the two greatest soldiers of the century, he writes:—

"He (the Duke of Wellington) said to me that he had always made it a rule to study for some hours every day; and alluded to his having commenced acting upon this rule before he went to

India, and of his having continued to act upon it. This is a fact that I apprehend is unknown as to the Duke of Wellington, and it is a very important one. It proves that, like Caesar and Napoleon, and probably all the greatest men of the world, he was fully aware of the necessity of systematic and careful study. It is evident that Napoleon, before he was twenty-six, had read all the works of all the great masters of war; had analysed their campaigns, and arranged fully in his mind the good and great principles which they established; and had thus filled his mind with such a body of principles, such a fund of knowledge, that no state of circumstances could be presented to him that he had not the means of solving. The rapidity and correctness of his solutions, under the most complicated circumstances, seemed mysterious to those who witnessed them; but that rapidity and correctness of solution unquestionably proceeded from that fund of principles established in his mind, aided by his wonderful natural powers."

Again, "Here we arrive at the same important result before alluded to, that great commanders, such as Caesar, Wellington, and Napoleon, having their minds fully prepared by the study of general principles, and combining that with some experience, solve on the instant, and in the midst of the greatest turmoils of war, its most difficult and important problems; while commanders, unprepared by the careful study of principles, feel themselves in constant doubt and difficulty, as new cases present themselves, requiring altered arrangements to meet them; and those, in war, must be constant."

"A soldier," said Frederick the Great, "should learn war before making it, and then strive to put his science into practice;" and the opinion of a veteran of the last century is a fitting corollary—"He who understands the theory of war would be better qualified for reducing the theory to practice than he who is deficient in such preliminary knowledge."

Mighty names have here been cited, but it will be unwise to argue that where there is no ambition to become a Caesar or a Napoleon, there is no need to follow their example in respect of theoretical study. The chances of war are such that an officer, even of the most subordinate rank, may find himself so placed, that on his grasp of the tactical requirements of the situation the issue of battle will depend.

In the event of England entering upon a great war it is happily evident that the regular army will be largely reinforced by contingents of home as well as of colonial volunteers. That the aid rendered by such contingents may be as efficient as possible at the earliest moment should be the aim of all who aspire to serve with them; and towards that efficiency the thorough training of their officers, commissioned and non-commissioned, will contribute much. To them this study is specially offered. Few of them have opportunity of gaining experience in the field; let them draw then on the experiences of others, and from them learn, as far as possible theoretically, how the manifold duties and requirements of active service are to be met and fulfilled.

The campaign of Fredericksburg has been selected, amongst other reasons, as having been fought by two armies very largely composed of unprofessional soldiers. The lessons it teaches, the shortcomings it reveals, are likely, therefore, to be of exceptional interest and value to that class of officers to whose consideration I venture to recommend them.

In the general account of any great action, it is seldom possible to narrate at length the action of so small a unit as a battalion, but company officers need not, on this account, be deterred from the study of operations of large masses of troops. The same tactical principles regulate the combat of a large force and of a small; and it is the thorough grasp of those principles, combined with courage and coolness, that makes a capable leader, whether of a section or an army corps.

I have not hesitated to relate in full various incidents of the conflict, nor to refer at length to the military virtues and vices of the American armies; and many details have been added in order to throw light upon the causes which affected the action of the opposing generals and of the troops they commanded.

In few lines does this sketch pretend to originality; it is, in fact, but an amplification of the chapter on Fredericksburg, in Colonel Chesney's "Campaigns in Virginia." I have borrowed right and left; and on any point which appears to invite comment, I have brought forward, as often as possible, the opinions of eminent military critics. I have striven to give an accurate relation of events, and of the condition and composition of the rival forces; and if I have been led away in any respect by partiality, I can only plead in excuse (looking to the issue of the war),

"Victrix causa Deis placuit, sed victa Catoni."

Lastly, I am not without hopes that this account of a brief yet remarkable campaign of the War of Secession may attract attention, more than has hitherto been the case, to the records of that mighty conflict; for they contain some of the noblest and most stirring passages in the military history of the English-speaking race, passages written in the blood of unprofessional soldiers.

THE CAMPAIGN OF FREDERICKSBURG

General view of the situation, and record of events from September 17th to November 30th.

On the 17th day of September, 1862, the Confederate Army of Northern Virginia, under General R. E. Lee, retiring from the abortive invasion of the State of Maryland, was attacked at Sharpsburg, a village on the Antietam Creek, two miles N. of the river Potomac, by the Federal Army of the Potomac, under General McClellan. [Map I.] The Federals numbered over 80,000, the Confederates under 40,000. At the close of the day Lee still held his ground, and had inflicted on his opponents a loss of 12,500 men.

During the 18th, the Confederates stood at bay, awaiting the attack which McClellan, with his demoralized host, dared not renew. During the night, without molestation, they crossed the river and regained Southern territory.

It will be sufficient for the purposes of this sketch to note that the Potomac formed the boundary between the hostile republics; that the objective of the Northern forces was Richmond, the capital of the State of Virginia, and the seat of the Confederate Government; and that between the Federals and that city stood Lee's army only, alone and unsupported.

The Southern forces retired leisurely to Winchester, a country town in the valley of the Shenandoah, west of the long and lofty range of the Blue Ridge. Not until the end of October did McClellan, though he claimed a decisive victory at the Antietam, cross the river at Berlin and cautiously follow his enemy, moving, however, on the eastern flank of the mountains.

While advancing on Warrenton, a small town on the Orange and Alexandria railway, he was superseded by President Lincoln, and replaced by General Burnside, who, though unwilling to accept the responsibility, was unfortunately for himself and the army unable to refuse it.

MAP I

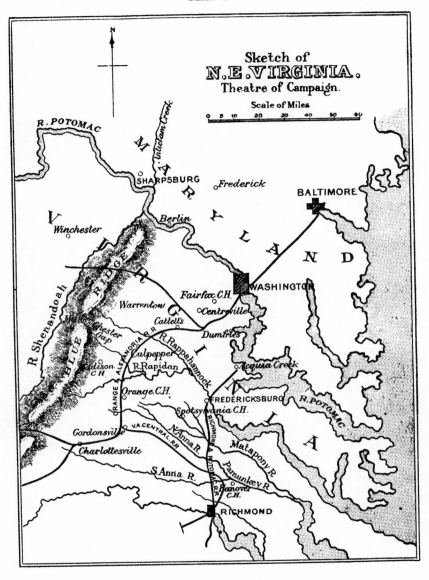

The opposing forces were at this date, November 9, disposed as follows:—

Federals, 125,000, moving south, at and about Warrenton; supplied from Alexandria.

Confederates, 75,000.
{ 1st Army Corps, under General Longstreet, in the neighbourhood of Culpeper, watching the passages of the Rappahannock; supplied from Gordonsville and Richmond.
2nd Army Corps, under General "Stonewall" Jackson, at Milford, twenty-nine miles due west of Warrenton, on the far side of the Blue Ridge; supplied locally.

The two Confederate wings were thus thirty miles, or two long marches, distant from each other, and separated by a range of mountains, which is pierced, however, by numerous easy passes or "gaps."

We may note that on the advance of the Federals to Warrenton the whole of that portion of Virginia which lies north of the Rappahannock passed into their hands, but the country was exhausted and the population hostile.

At this juncture Burnside ordered a halt, and remained quiescent for a week, occupying himself in simplifying the organization of his army, in drawing up a plan of operations, and in making provision through the War Department at Washington, the Federal capital, for a change of base.

McClellan had intended to crush the divided forces of the Confederates in detail, moving first against Longstreet; his successor, influenced by the presence of the dreaded Jackson behind the mountains, menacing his communications, did not consider the scheme feasible, and made no attempt to put it to the test.

Two routes to Richmond through northern Virginia lay before him: that of the Orange and Alexandria railroad, on which McClellan had intended to move, and that of the Richmond and Potomac line, which touched Union territory at Aquia Creek on the Potomac, nine miles N.E. of the city of Fredericksburg. On this latter Burnside decided, and with a view to immediate operations made arrangements with the Washington authorities to furnish him with a pontoon train, and to form a large supply depôt at Aquia Creek.

The line Fredericksburg-Richmond possessed several advantages. It was only sixty miles in length, far shorter than the

Gordonsville route; and to adopt it would cover Washington, a fact which commended it to the President and his Cabinet; for, if Lee were to make a dash at the northern capital, Burnside's army could be transferred thither by way of Aquia Creek and the river with great ease and rapidity. This anxiety of Lincoln and his advisers for the security of Washington prevented the adoption of a base of invasion south of the James river, by which Richmond was eventually captured, until the year 1864; this alternative has not therefore been mentioned as open to Burnside.

On November 15th, the Federal army began its march, moving in a south-easterly direction along the north bank of the Rappahannock towards Falmouth, a village thirty-five miles distant from Washington, and two miles above Fredericksburg, which lies on the opposite shore. [Map II.]

Of this change of direction Lee was at once informed by his cavalry scouts. On the 19th, divining his opponent's purpose, and ignoring the forward movement of a hostile detachment towards Culpeper—a ruse by which Burnside had hoped to distract his attention and conceal his real design—he ordered Longstreet's corps to march on Fredericksburg, on a line parallel to that taken by Burnside, the Rappahannock and Rapidan flowing between them.

The action at this period of the Southern Cavalry (four brigades under General J. E. B. Stuart) is well worthy of notice.

With a total strength (on paper) of 9,000 sabres, this arm watched a line extending along the Rappahannock and bending as far south on either flank as Culpeper and Spotsylvania Court House, a distance of fifty miles (an average of a hundred and eighty men to the mile); and at the same time made frequent and extended reconnaissances.

On November 15th, a patrol discovered that Federal troops were moving eastward, and another on the same day brought news that gunboats and transports had entered Aquia Creek on the Potomac. These two pieces of information, collected at points nearly forty miles distant from each other, gave General Lee an inkling as to his opponent's design; and the small infantry garrison which had hitherto held Fredericksburg was consequently reinforced by a strong detachment of cavalry and a field battery.

On the 16th two supply-trains were captured near Warrenton. On the 17th a party of Confederate troopers rode into Catlett's Station, fifteen miles N. of the Rappahannock, an hour or so after the wagon-train of the Federal advanced guard had left it.

On the 18th, Stuart in person, with one regiment and three horse artillery guns, forced the passage of the river near Warrenton Springs, and discovered, in the town of Warrenton, that half of Burnside's army had already passed through *en route* for Falmouth. The next day Longstreet was despatched to Fredericksburg.

The Federal General Sumner, with 33,000 men, reached Falmouth on the afternoon of the 17th, having compassed thirty-five miles in two days and a half. Fredericksburg, two miles below on the opposite bank of the river, was at this time occupied only by four companies of Confederate infantry, a regiment of cavalry, and a light battery. Sumner, finding the river to be fordable in the vicinity, wrote to Burnside, on the night of his arrival, for instructions as to whether he should occupy the town and the heights beyond on the following morning.

The commanding general replied that he did not think such a movement advisable until his communications with Aquia Creek had been established.

At this time the pontoons demanded on November 13th, owing to the culpable mismanagement of the Washington authorities, had not arrived; the river, swollen with recent rains, was rising fast; one of the fords was but a series of submerged stepping-stones; no large quantity of supplies had been as yet collected at Aquia Creek; the road between that point and Falmouth was in very bad condition; the railway had been destroyed, and neither wharves nor landing-place existed.

On the 19th and 20th, the remainder of the Federal army arrived at Falmouth, and 110,000 men were concentrated within a few miles of Fredericksburg.

General Sigel, with Eleventh Federal Corps, had been already detached, and remained near Centreville, covering Washington. To General Banks and an army of 20,000 men was entrusted the duty of guarding the Upper Potomac and the issues from the Shenandoah valley. [Map I.] These two officers were not under

Burnside's orders, nor did General Banks's troops form part of the Army of the Potomac.

On the 21st, Longstreet, with two divisions, the advanced guard of the Confederate right wing, arrived at Fredericksburg, and taking post on the heights commanding the town, began immediately the construction of entrenchments.[1]

On the 22nd, his three remaining divisions, two of which, Hood's and Pickett's, had marched by way of Spotsylvania Court House, moved into line. The corps was 35,800 strong.

General Lee, although aware that the whole Federal force was threatening Fredericksburg, was in no haste to reinforce his right. The season was advanced; the rigorous Virginian winter was close at hand, when protracted operations would be impossible; and as well-informed, probably, as Burnside himself of the Federal difficulties as regards supply, he was so little apprehensive of immediate attack, that not until the 22nd did he summon Jackson from the valley of the Shenandoah.[2] Crossing the Blue Ridge by Fisher's Gap, that general brought up the left wing by way of Madison Court House, and, reporting to the commander-in-chief on the last day of the month, took up his position to Longstreet's right and rear.

Although the pontoon train had arrived by the 24th, during this period the Federals had made no aggressive movement. A wanton threat of bombarding Fredericksburg had caused the evacuation of the town by the greater part of the inhabitants, who, exposed without shelter to the bitter November weather, endured much unmerited suffering.

Thanks to the energy of the Federal Engineer and Departmental officers, the railway between Falmouth and the Potomac was restored and in working order by the 28th; at Aquia Creek supplies sufficient for at least a fortnight's consumption with much *matériel* had been accumulated, and a commodious har-

---

1. [In his report on the battle of Fredericksburg, Longstreet stated that he reached Fredericksburg on November 19. McLaws' division arrived on the twentieth, followed the next day by the remainder of the corps. Wilcox's brigade of Anderson's division did not arrive until November 23 (*Official Records*, Ser. 1, XXI, 568, 584, 610).]

2. [Lee's dispatch to Jackson requesting him to move east of the Blue Ridge was dated November 23. Jackson's corps had passed through Fisher's Gap by the twenty-fifth and the following day was ordered to join Lee at Fredericksburg (*ibid.*, pp. 1027, 1033).]

bour and substantial quays constructed. Sixteen days only had elapsed since the first orders for these services had been received.

Lee certainly acted boldly when he pushed his right wing so far forward, and placed it within the angle formed by the Rappahannock above Fredericksburg; and it may seem at first sight that he acted rashly, considering that the river could be crossed at any point by an enemy possessing a pontoon train, and Longstreet's position be turned on either flank. It was not so, however: we shall find on examination that his dispositions were perfectly sound, and adequately provided for every emergency.

His main object was to prevent the Federals, if possible, from occupying Virginian soil south of the Rappahannock. Apart from the moral effect of yielding to such aggression, the country still, though with difficulty, maintained a scanty population which he was unwilling to expose to further hardships.

To meet Burnside, if he advanced, as near the Rappahannock as might be, was, with this object in view, his principal aim; and he therefore assumed a position on the bank of that river. He was convinced, as we have seen, that no advance could be made for some time, and was therefore in no hurry to call up Jackson; but nevertheless his precautions were taken against a sudden attack. Every ford was guarded, scouts continually penetrated the enemy's lines, and much valuable information was received from the Virginian citizens who still remained in their homes beyond the Rappahannock.

The ridge in rear of Fredericksburg, occupied by Longstreet, was a formidable defensive position, and commanding as it did the town and the river beyond, rendered the task of crossing there no easy one.

Jackson, moving leisurely towards the Rappahannock by way of Madison and Orange Court Houses, which latter he reached on November 26th, was prepared, if Burnside had moved to his right and crossed the river at Banks's and United States Fords, to unite with Longstreet at Chancellorsville or Spotsylvania Court House; or, being everywhere within easy reach of the railway, to join him near Hanover Court House on the North Anna, a strong line of defence, should the Federals succeed in crossing at Fredericksburg or below. [Maps I, II.]

Orange Court House is twenty-four miles equidistant from Chancellorsville and Spotsylvania Court House; *via* United States Ford, Chancellorsville is only twelve miles, and Spotsylvania Court House twenty, from Falmouth, but the Rappahannock is between.

In short, Lee possessed all the advantages of interior lines, and relying upon the watchfulness and enterprise of Stuart and his horsemen, he knew that he would receive long warning of any projected movement, turning or otherwise, and have ample time to concentrate his forces to meet it.

Disposition of troops during first week of December.

At the beginning of December, Burnside, urged on by the clamour of the northern press and populace, resolved to cross the Rappahannock, and despite the near approach of winter to assume the offensive. On the 5th, he moved down stream from Falmouth, and attempted to surprise a passage at Skinker's Neck, a bend of the river, twelve miles below Fredericksburg. Lee, however, having learnt or anticipated his intention, had already, two days before, sent D. H. Hill's division of Jackson's Corps to Port Royal, a few miles below the threatened point; and Burnside, finding the enemy prepared to resist his crossing, abandoned the enterprise. Two attempts of Federal gun-boats to pass up the river about this time were frustrated by Stuart's horse artillery and some field batteries. Immediately on his arrival in the vicinity of Skinker's Neck, General Hill had reported that the locality presented many advantages to a force attempting the passage from the left bank, and also that he had discovered indications that such a movement was in contemplation. Early's division of Jackson's Corps had thereupon been sent to his support; and the high ground above the river, between Skinker's Neck and Port Royal, was prepared for defence by these troops.

The Confederate forces were at this time thus disposed [Map II]:—

1st Corps (Longstreet) { Held the ridge in rear of Fredericksburg.

2nd Corps (Jackson)
- 1 Division (A. P. Hill) near Yerby's House, five miles S. of Fredericksburg.
- 1 Division (Taliaferro) at Guiney's Station, on the railroad, nine miles S. of Fredericksburg.
- 1 Division (Early) at the Hop Yard, on the Rappahannock, twelve miles S. of Fredericksburg.
- 1 Division (D. H. Hill) at Port Royal, on the Rappahannock, eighteen miles S. of Fredericksburg.

Cavalry (Stuart)
- 1 Brigade (Rosser) in Spotsylvania County.
- 1 Brigade (W. H. F. Lee) at Port Royal.
- 1 Brigade (Hampton) on the Upper Rappahannock.
- 1 Brigade (Fitzhugh Lee) with Longstreet.

This distribution is an excellent example of the most judicious method of guarding a line of river, *i.e.*, a series of strong detachments so placed as to be able to concentrate rapidly and to anticipate the enemy at any threatened point.

The Federal army was encamped round Falmouth, with the exception of a small force retained for purposes of demonstration near Skinker's Neck. The bank of the stream was constantly patrolled, and the Confederate movements observed as far as possible. Reconnoitring balloons were brought into use, but no scouting parties crossed the stream; and behind the thick forest which shrouds the valley of the Rappahannock the enemy's camps and dispositions were completely screened.

It may here be noted that such was the moral effect of General Lee's invasion of Union territory, earlier in the year, that the Federal Government, though aware that the whole of the Confederate army was massed near Fredericksburg, dared not set Banks and Sigel free to strengthen Burnside. Thus during the campaign, more than 35,000 Federal troops were standing idle, detained at a distance to guard against an attack on Washington, which the Cabinet feared might be executed by an enemy who was already confronted by superior forces!

The Confederates acted with greater boldness and better judgment. Between Lee's army and Richmond stood not a single brigade.

Before proceeding to the further operations it will be as well to give a few details of the organization, quality, and composition of either army, and a brief description of the theatre of the campaign.

MAP II[3]

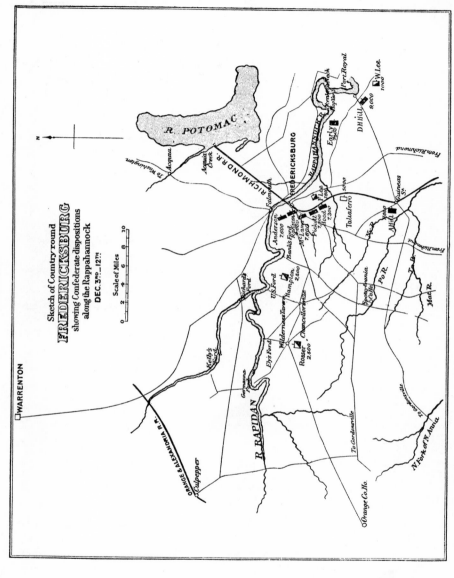

3. A. P. Hill's division should be located at Yerby's House and Taliaferro's division at Guiney's Station.

THE ARMY OF THE POTOMAC.

The Federal army had been distributed by Burnside into three Grand Divisions, and on November 10th the numbers present and fit for duty were shown as follows:—

THE RIGHT GRAND DIVISION.
MAJOR-GENERAL SUMNER.
Infantry and Artillery, 22,736.

II. Corps (Major-Gen. Couch)	1st Division (Hancock)	1st Brigade (Caldwell). 2nd ("Irish") Brigade (Meagher). 3rd Brigade (Zook), 2 Batteries.
	2nd Division (Howard)	3 Brigades and 2 Batteries.
	3rd Division (French)	Ditto.
IX. Corps (Major-Gen. Willcox)	1st Division (Burns)	2 [3] Brigades and 2 Batteries.
	2nd Division (Sturgis)	Ditto. [2 Brigades and 4 Batteries]
	3rd Division (Getty)	Ditto. [2 Brigades and 2 Batteries]

Cavalry.

THE CENTRE GRAND DIVISION.
MAJOR-GENERAL HOOKER.
Infantry and Artillery, 39,984.

III. Corps (Major-Gen. Stoneman)	1st Division (Birney)	3 Brigades and 4 [2] Batteries.
	2nd Division (Sickles)	3 Brigades and 3 [4] Batteries.
	3rd Division (Whipple)	2 Brigades and 3 Batteries.
V. Corps (Major-Gen. Butterfield)	1st Division (Griffin)	3 Brigades and 2 [4] Batteries.
	2nd Division (Sykes)	Ditto. (18 regiments [including 12] of regular infantry U.S.A. [and 2 Batteries]).
	3rd Division (Humphreys)	2 Brigades and 2 Batteries.

Cavalry.

LEFT GRAND DIVISION.

MAJOR-GENERAL FRANKLIN.

Infantry and Artillery, 42,892.

I. Corps (Major-Gen. Reynolds)
- 1st Division (Gibbon)[Doubleday] — 3 [4] Brigades and 4 [3] Batteries.
- 2nd Division (Meade)[Gibbon] — Ditto. [3 Brigades and 4 Batteries]
- 3rd Division (Doubleday)[Meade] — Ditto. [3 Brigades and 4 Batteries]

VI. Corps (Major-Gen. Smith)
- 1st Division (Brooks) — Ditto. [3 Brigades and 4 Batteries]
- 2nd Division (Howe) — Ditto. [3 Brigades and 4 Batteries]
- 3rd Division (Newton) — Ditto. [3 Brigades and 3 Batteries]

Cavalry.

Chief of Artillery (Major-Gen. Hunt) } Artillery
- Sumner 60
- Hooker 100
- Franklin 116

Field guns.

Heavy (about) 74
—————
350

Grand total, 109,612 men, and 350 guns.

ARMY OF NORTHERN VIRGINIA.

Present for duty December 10, 1862.

I. Corps (Lieut.-Gen. Longstreet)
- 1st Division (Hood) — 5 [4] Brigades.
- 2nd Division (Pickett) — 5 Brigades.
- 3rd Division (Ransom) — 2 Brigades { Cooke. Ransom. }
- 4th Division (McLaws) — 4 Brigades { Cobb ("Irish"). Kershaw. Barksdale. Semmes. }
- 5th Division (Anderson) — 5 Brigades.

Total 35,806.

Reserve Artillery { (Col. Walton). (Col. Alexander). }

The Campaign of Fredericksburg 25

II. Corps (Lieut.-Gen. Jackson)	1st ("The Light") Division (A. P. Hill)	6 Brigades	Gregg. Archer. Field. Lane. Pender. Thomas.
	2nd Division (D. H. Hill)	[5 Brigades]	
	3rd Division (Early)	4 Brigades	Walker. Atkinson. Hoke. Hays ("Irish").
	4th ("The Stonewall") Division (Taliaferro)	4 Brigades	Paxton. Jones. Warren. Branch. [Starke]

Total 33,595.
Reserve Artillery (Col. Walker).

Cavalry and Horse Artillery.

(Maj.-Gen. Stuart)
1st Brigade (Rosser).
2nd Brigade (W. H. F. Lee).
3rd Brigade (Hampton).
4th Brigade (F. Lee).
Total 9,114.

Chief of Artillery (Brigadier-Gen. Pendleton)   *Heavy Artillery.*
Total 713.
Grand total, 78,228 men, and probably 250 guns, including a complement of heavy cannon.

In both armies the strength of divisions varied from 4,000 to 12,000; while the average muster of rank and file in infantry battalions was about 350. Four to six guns (often of different calibre) composed a battery; and the Confederate cavalry brigade numbered 2,000 sabres.

Between the military forces of the North and South existed, as we should expect to find, a strong family likeness. They were based upon the same military system, were raised at the same moment and in the same manner, and the same causes forbade either assuming the distinctive features of a regular army. It will be instructive to observe the characteristics common to both, and at the same time certain essential differences which bore decisively upon the issue of their campaigns.

In the Federal army the officers of the superior grades had, almost without exception, served throughout the war, and in three arduous campaigns had won experience. Few of those, however, below the rank of field-officer had any higher military qualifications than zeal and courage. Want of previous training and the fact that they owed their commissions to the vote of the rank and file and not to the selection of the War Department had to answer for their shortcomings. The choice of the privates did not always fall upon the most efficient soldiers, but more often upon those whom self-assertion, a reputation for daring, or even social qualities made the heroes of the hour. Again, many who would have worthily filled the higher rank were unwilling to take up the responsibility of command, and the invidious task of administering discipline on those who had so lately been their equals. A system of examination, instituted by McClellan, had weeded out many of the hard bargains; but among the company officers the standard of discipline and efficiency was still very low. The same remarks apply to the non-commissioned ranks.[4]

The private soldiers comprised men of every grade of society and of many nationalities. Men of substance and education, the best breeding and culture of the North, marched and fought shoulder to shoulder with labourers and mechanics, with *gamins* from New York and half-civilized immigrants. The Federal hosts were not recruited from one continent alone. The proportion of American-born was much larger than has been generally supposed, but the speech of every European nation might have

4. [Most European observers who visited the Union army during the first two years of the war agreed that the standard of efficiency among company officers was low. The Marquis of Hartington inspected a regiment of New York Volunteers in 1862 and wrote home that it "seems a great pity that such fine material should be thrown away . . . by having utterly incompetent officers" (Bernard Holland, *The Life of Spencer Compton Eighth Duke of Devonshire* [London, 1911], I, 44). Captain Edward Hewett, who likewise visited the Union army in the fall of 1862, blamed most of the faults of the soldiers on incompetent leadership. Unless better officers were developed soon, Hewett wrote, all the arms and equipment "of these insane armies will be unserviceable in a year or so" (R. A. Preston [ed.], "A Letter from a British Military Observer of the American Civil War," *Military Affairs*, XVI [Sumner, 1952], 53). Lt. Col. Ferdinand Lecomte, the Swiss military observer who accompanied McClellan's army during the Peninsular campaign, believed that the election of officers by the rank and file produced mediocre leadership (*The War in the United States: Report to the Swiss Military Department* . . . [New York, 1863], pp. 93–94).]

been heard in the camps of the Army of the Potomac. There were brigades of Irish, famous for their desperate courage, and more than one division of Germans. There were some who had worn the red shirt with Garibaldi, and others who had rebelled with Kossuth. Canada sent many gallant riflemen, nor were men wanting who had fought beneath the Tricolor or Union Jack. The unfurling of the Stars and Stripes had gathered together a motley assemblage. Attracted by honour, gold, or sheer love of the soldier's trade, the patriot and the hireling, the citizen and the alien, the prince and the adventurer, stood beneath its folds; and out of this heterogeneous mass the able administration of McClellan had welded a highly organized and efficient force.

The spirit of the troops was good, and although during its eighteen months of existence the Army of the Potomac had met with many reverses and but few successes, its *élan* and confidence was still unbroken. The *morale* of such a force, leavened by the presence of men of intelligence and high principle, was necessarily good. Crime was practically unknown. Public opinion was more dreaded than the provost-marshal, and ever ready to condemn, proved a salutary check on all serious misbehaviour. At the same time the more trifling offences were overlooked. The dogma of absolute equality interfered with the discipline of the republican soldiery, and that habit of prompt obedience was wanting, without which no army can prove a satisfactory instrument in the hands of its commander. The general though tacit consensus of the privates defined the limits of authority, and no superior dared demand further subordination than they thought reasonable to yield.

Nor did the regimental officers possess the qualifications necessary always, but more especially in an army where all were socially equal and where the claims of rank were grudgingly allowed, to secure the trust and obedience of the men. Superior knowledge of the art of war, thorough acquaintance with duty and large experience, seldom fail to command submission and respect. Lacking these qualifications, the Federal officers had but little hold upon their men; and the soldier was wont to regulate his action rather by the opinion of his comrades or by his own judgment than by the voice of his superior.

Later in the war, when General Grant took over the chief command, this state of things had improved materially; but by that time the officers were far more competent to command; and the men, having at length recognized its utility, did not resist the introduction of a somewhat stricter code of discipline.

It may be mentioned that though the patriotism of the majority was pure and exalted and their motives for enlisting most worthy, still there was a large element which had been attracted by the sordid allurements of ample bounties, and whose spirit and resolution were therefore much inferior to those of men who were animated by loftier sentiments.

The Federal artillery was numerous, and the weapons of the best and newest patterns, including a large proportion of rifled guns; but General Hunt, Chief of Artillery in the Army of the Potomac, has testified that the officers of this arm, who were most of them drawn from the regular service, had received a thorough technical but no tactical training whatever, and were ignorant of the value and capabilities of their own weapons.

The cavalry was inferior: the troopers well mounted but indifferent horsemen, and their leaders had hitherto shown but little enterprise.

The engineers were efficient, for the large numbers of civil engineers who had enlisted, men used to the rough exigencies of a new country, required but little training to become excellent officers.

The infantry soldiers, as a rule marched fairly well, were brave and stubborn in the field, and patient under reverses; armed with a serviceable and long-ranging rifled muzzle-loader, they were but indifferent marksmen, and often extremely careless of the condition of their arms.

The staff was in fair working order and full of zeal; but to the scarcity of professional soldiers in the appointments of lower grade many of the Federal disasters were certainly due. Among the corps and division leaders there was great lack of accord and no proper system of subordination. The major-generals in command of Grand Divisions communicated directly with the military authorities at Washington, without reference to their immediate superior. So little were the claims of discipline acknowledged or understood even in the highest ranks, that Hooker, commanding the Centre Grand Division, before the battle of Fredericksburg, did not hesitate to openly express his disapproval of General Burnside's plans and his forebodings as to the result: these opinions spreading rapidly through his command, sapped the confidence of the soldiers, and sent them into action half-beaten before a shot was fired. So gross an indiscretion has few parallels in history.

The Ordnance, Transport, and Supply departments, though the knavery of the contractors has become a byword, were excellently organized and served. The armament and equipment of the troops left little to be desired, and the medical and hospital machinery was well nigh faultless. The vast wealth and resources of the Union Government were unsparingly lavished on the comfort and well-being of the soldiers; and while none, perhaps, have ever been so well cared for as those of the North, few, on the other hand, have endured greater hardships than did the men who formed the Army of [Northern] Virginia.

One division of regular infantry and some batteries of the old United States army were with the Army of the Potomac, but the remainder of the troops were volunteers, that is, they had received no regular military training, and had only adopted the profession of arms as a temporary measure.

The Confederate army was also, with the exception of many officers, composed entirely of unprofessional soldiers; and in organization, composition, and discipline, had much in common with its rival. The same mixture of classes existed in the ranks; but there were few foreigners. Discipline was perhaps less strict, and how slight were its bonds the following extracts show. In a work on the inner life of this very Army of Northern Virginia, written by one who had served in it, it is said that "it took years to teach the educated privates that it was their duty to give unquestioning obedience to officers because they were such;" and that "the conflict (between officers and men) was soon commenced and maintained to the end."

We read elsewhere that "at the drawn battle of Sharpsburg, September 17, 1862, where Lee so successfully resisted the onslaught of a far superior force, 25,000 stragglers, more than one-third of his numbers, were absent from his ranks."

That eminent writer, Colonel Charles Chesney, in speaking of

Lee's errors, charges him as follows:—"Chief of these was his permitting the continuance of the laxity of discipline which throughout the war clogged the movements of the Confederates, and robbed their most brilliant victories of their reward. The fatal habit of straggling from the ranks on the least pretext; the hardly less fatal habit of allowing each man to load himself with any superfluous arms or clothes he chose to carry; the general want of subordination to trifling orders, which was the inheritance of their volunteer origin,—these evils Lee found in full existence when he took command before Richmond, and he never strove to check them. Nor did he use his great authority, as he might have done, to purge his command of the many inefficient officers whose example of itself was ruinous to all discipline. . . . As the war went on the rifts caused by indiscipline and carelessness in the Confederate armour widened more and more; and in the end those faults were hardly less fatal to the South than the greater material forces of her adversary. Her fall was to offer new proof to the world that neither personal courage nor heroic leadership can any more supply the place of discipline to a national force, than can untrained patriotism or the vaunt of past glories."

In numbers the Southern army was much inferior. The deficiency in quantity, however, was fully compensated for by superiority in quality. The Confederate soldiers were drawn from the farm and the plantation, for there were few large towns in the seceded States. They were consequently trained from childhood to the use of fire-arms, excellent marksmen, practised horsemen, and of hardy constitution. Habituated to a life of sport and adventure in the wilderness and forests of the South, it was easier for them to become soldiers, and they were more reliable in the emergencies of battle than their city-bred opponents.

No country ever possessed finer material for soldiers than the Confederacy, and it has been conceded on all hands, by none more readily than those who met it so often and proved such worthy foes, that the Army of Northern Virginia was, on a fair field, absolutely invincible.[5] To say, as we must, that the soldiers

---

5. "Who can forget, that once looked upon it," says a Northern writer, "that array of tattered uniforms and bright muskets, that body of incomparable infantry, the Army

of the Army of the Potomac were inferior only to Lee's veterans is to award them the highest praise.

From the first battle of Bull Run, fought in July, 1861, the Confederate forces in Virginia had gone on from victory to victory; for even indecisive battles, owing to the odds encountered and the terrible loss inflicted, had been moral successes. With unbounded faith in their leaders and well-warranted confidence in their own powers, Lee's battalions feared no weight of numbers their gigantic foe might bring against them. Reciprocating their trust, the great American captain never hesitated to engage, however overwhelming the odds that faced him.

The antipathy of a landed aristocracy to a radical democracy, the deep resentment of a proud minority at what it deemed the arrogance of an unscrupulous majority, nerved the arm and steeled the heart of the Confederate soldier. Loyalty to that young nation which the Ordinance of Secession had called into being, and an earnest belief in the rectitude of his cause, took the place of the broader and more practical patriotism of the North. Nor did the knowledge that the homes of the South were being wasted, and that there was no hope of freedom and independence save in victory alone, fail to intensify his resolution.

The Confederate Civil departments were as inferior in every respect to the Federal as the resources of the South in comparison with those of her rival were scanty; and moreover the organization and management of these indispensable branches were exceedingly faulty.[6] Armament and equipment were good, for they were the spoils of many victories.

The personnel of the artillery was efficient; the guns numerous, but inferior in quality, and the fuzes exceedingly bad.

The cavalry, or rather the mounted infantry—for the troopers, armed with rifles and carbines, fought as a rule on foot—did excellent service on the outpost line; and the extended raids and

---

of Northern Virginia, which, for four years, carried the revolt on its bayonets, opposing a constant front to the mighty concentration of power brought against it, which, receiving terrible blows, did not fail to give the like, and which, vital in all its parts, died only with its annihilation?"

6. Lee, on December the 1st, reported to the Secretary for War that several thousands of his soldiers were without boots. [On December 2, Lee wrote that "there is still a great want of shoes in the army, between 2,000 and 3,000 men being at present barefooted" (*Official Records*, XXI, 1041).]

reconnaissances so successfully carried out by Stuart and his brigadiers are models of enterprise and daring. German soldiers have allowed that the lessons of the American War were not sufficiently taken to heart, and that the Uhlans in 1870–71, compared with Stuart's and Sheridan's audacious horsemen, were timid and unenterprising.

The senior officers of the Confederate staff were able and energetic; but the want of trained aides-de-camp was severely felt. The generals of corps and divisions were, without exception, devoted to and in full accord with their great leader, Lee.

Between the superior officers of the armies who met at Fredericksburg there can be no comparison. Lee and his famous Lieutenant, "Stonewall" Jackson, were without doubt the greatest commanders the war had yet produced, and with the intuitive perception of master minds they had gathered round them a group of splendid soldiers. On the Federal side, Meade, Hancock, and Humphreys were gallant and capable officers, and afterwards on many battlefields won well-deserved renown; during the Fredericksburg campaign they were but divisional generals, and were prevented, by blundering superiors, from displaying their undoubted talents.

One important fact, which materially affected the issue of the campaign before us, may here be noticed. The white population of the South was seven millions, of the North, twenty. The supply, therefore, of men for the Confederate armies fell far short of that commanded by the Union Government; and to increase the disproportion, the tide of immigration, always setting towards the Northern States, produced an endless harvest of recruits, from which source of plenty the Confederacy was debarred. The resources of the North in men, as in all else, were practically limitless. The South foresaw, if circumstances did not intervene, that her very manhood would be ultimately exhausted.

Looking at the fact that the United States had been engaged in no conflict since the Mexican war of 1847, if we except the warfare of the Indian border, it might fairly be asked, where did the superior officers of the Federal and Confederate armies gain that knowledge of the military art which many of them undoubtedly possessed? The answer is simple. The great majority had served in the regular army, and had been educated at the Military Academy of West Point, where the training is as complete as the discipline is severe. As showing the value of a sound military education, the fact is worth noting that, with few exceptions, no man made his mark in the Civil War who had not passed through the four years' course of the military school. The volunteer generals had every chance offered them, and in the North were specially favoured; but, though many made good brigadiers, few proved themselves fit for more independent command. At Lexington, in Virginia, an academy, formed on the model of West Point, had existed for some years before the war. There the famous Jackson had been a professor, and there many of the Southern officers had received a semi-military training.

Nor had the experience of the protracted Mexican campaigns been thrown away on the American army. Many of the generals on either side had served in them: there the genius of Lee, the daring of Jackson, had first won notice; there Grant and Beauregard, McClellan and Johnston, had learnt their earliest lessons in civilized warfare.

Burnside was a "West Pointer," but had retired from the army after six years' service. The circumstance that the rank and file of the regular army had adhered to the Union colours was not altogether a fortunate one for the North. Many of the officers continued to serve with their own regiments; while in the Confederacy, those who had "come South" were placed in command of brigades and regiments or distributed on the staff, and their wholesome influence was thus felt by the army at large, and not confined to any one corps or division.

The Confederate infantry, with the exception of some picked divisions, did not excel in marching, although in neither army would the soldiers submit to carry the heavy equipment of regular troops. Nor was discipline strong enough to prevent the men wasting their rations. During the arduous operations the Confederate troops were often starving and exhausted at the critical moment, and their ranks decimated by stragglers in search of food. This state of things was due to the incapacity of the regimental officers and the neglect of their superiors to tighten the bonds of discipline.

The drill and formations of both armies were identical. Infantry regiments were composed of one battalion of ten companies.[7] The movements were few and simple. Column formations were seldom employed except on the line of march. For attack and defence eight companies were drawn up in "line of battle" of two ranks, covered by the remaining two as skirmishers; these latter closed to the flank as the line approached the enemy, and joined their regiments in the assault. Deployments were effected, and the skirmish line thrown out, with rapidity and precision. On the line of march, little attention was paid to such necessary details as dressing and distance, and straggling was looked upon as but a venial offence.[8]

The wooded nature of the country over which they fought had taught the intelligent American soldier the value of entrenchments, and the troops were adepts in extemporizing useful defences.[9] Without waiting for orders they were accustomed, whenever they took up even a temporary position, with axe, spade, and bayonet, to cover themselves from fire.

The cavalry of both armies was trained as mounted infantry, and at this period relied far more on the rifle than on *l'arme blanche*.[10] This mode of fighting, like the use of entrenchments, was made compulsory by the nature of the country, affording as

7. [American readers may be puzzled by Henderson's use of the word "battalion." After the Cardwell Reforms in the 1870's, the practical working unit of the British infantry was a battalion of eight companies, with a war establishment of 1,096 officers and men (General Viscount Wolseley, *The Soldier's Pocket-Book for Field Service* [London, 1886], p. 28). The regiment was an administrative unit only: after 1881 the typical regiment comprised "two Line battalions, two militia battalions, the regimental depôt, and the volunteer battalions existing within the area of the district" (Captain Owen Wheeler, *The War Office Past and Present* [London, 1914], pp. 231–32). In contrast, during the spring following the battle of Fredericksburg, the average regiment of infantry in the Army of the Potomac numbered about 433 and in the Army of Northern Virginia about 409 effectives (John Bigelow, Jr., *The Campaign of Chancellorsville: A Strategic and Tactical Study* [New Haven, 1910], p. 27).]

8. Thorough discipline alone assures orderly marching; without orderly marching the exact execution of the leader's combination is difficult, and the timely presence of superior numbers at the decisive point improbable.

9. A soldier of Lee's army has recorded that "experienced men in battle always availed themselves of any shelter within reach—a tree, a fence, a mound of earth, a ditch, anything . . . only recruits and fools neglected the smallest shelter."

10. [*L'arme blanche* literally means side arm, or cutting weapon. In its usual context it refers to the traditional cavalry shock tactics of charging in massed formation with drawn swords (see below, pp. 124–25).]

it did but few facilities for cavalry manoeuvres on a large scale.

The field artillery of either army was attached to the infantry divisions. In the Confederate army artillery battalions existed, which formed a reserve and acted independently of the divisional batteries.

Nature and resources of the theatre of the Campaign.

That portion of Virginia lying east of the Blue Ridge [Map I], which was the theatre of this and many more campaigns, is covered with primaeval forest. Clearings are few and far between; the roads are bad, and the undergrowth of the woods is penetrated with difficulty. The face of the country is undulating, with gentle gradients, and abrupt or considerable eminences are seldom met with.

The water-courses are numerous, and several attain to the dignity of rivers, notably the Rappahannock and the Rapidan, the Mattaponi and the North and South Annas. Their course is, generally speaking, from west to east, crossing at right angles the line of Federal invasion.

The State is in many parts exceedingly fertile, but sparsely cultivated and without manufacturing industry or appliances; and at this period neither food nor forage were plentiful. It is still thinly populated, and between Richmond and the Potomac there is no large town or market.

The railroads have already been remarked.

The Rappahannock is tidal as far as Fredericksburg, and two hundred yards wide in the neighborhood of the town, increasing in width as it flows seaward. Below Port Royal it becomes a formidable obstacle in the path of a hostile army. There are several easy fords above Falmouth; the river-banks, except near Fredericksburg, are clad with timber, and from Falmouth downwards the left commands the right.

Burnside's object and plan.

Burnside's objective was, as we have seen, Richmond, sixty miles south of Fredericksburg; but between him and the Confederate capital intervened the broad waters of the Rappahannock and the Army of Northern Virginia. Before the city could be captured, Lee's army must be annihilated; before that army could be dealt with, the Rappahannock must be crossed.

He had therefore two distinct ends to accomplish: the first, to convey his force across the river; the second, to bring the Confederates to bay.

The first operation he had already attempted, and had failed; but in no wise disheartened by his ill-success at Skinker's Neck on December 5th, he devoted the following days to maturing a fresh scheme.

We will now consider what information he had as to the strength and disposition of Lee's troops—that is, what data he possessed on which to frame his plans; and also the course of action he ultimately adopted.

In the first place, it appears that he knew that Lee's whole army was on the line of the Rappahannock; and that, over-estimating somewhat the numbers at his adversary's disposal, he believed himself confronted by a force little inferior in numbers to his own.

Secondly, from the plateau of Stafford County, which crowns the left bank of the river, large Confederate camps were visible near the Hop Yard and Port Royal, and there were signs of the presence of a considerable force of infantry and artillery on the ridge in rear of Fredericksburg.

This was the extent of his information; and although between the Confederate position at Fredericksburg and that near the Hop Yard twelve miles of wooded upland intervened, no attempt was made by him or his cavalry leaders to cross the river and penetrate this screen, or to ascertain the strength and whereabouts of each hostile division. He seems to have assumed that half of Lee's army was in the vicinity of Port Royal, the remainder behind Fredericksburg. He neither knew nor suspected that two strong supporting divisions (A. P. Hill and Taliaferro) were concealed within the recesses of the forest within easy reach of either point.

Fredericksburg, at the head of the railroad, and in close communication with his base at Aquia Creek, was the spot where Burnside would have preferred to cross the river. Unfortunately the ridge behind the town, occupied by the enemy and partially prepared for defence, formed a very formidable position. It was evident that even if it were held by a small force only, loss would be incurred in forcing the passage. Held by the whole of Lee's army it would be almost impregnable.

At Skinker's Neck [Map II], twelve miles lower down, Burnside had been already foiled. The river there was broad and deep, the enemy in force and entrenched. To cross midway between Fredericksburg and this last named locality was scarcely practicable. The right bank was so carefully patrolled by Stuart's horsemen that surprise was out of the question. The first alarm would have drawn in the Confederate wings to occupy a strong line of defence along the forest-clad heights.

To operate below Port Royal was equally impracticable. The increased width of the river was a serious obstacle; and as Washington would have been unmasked by the movement, the President would have certainly vetoed any such scheme.

There was one last course feasible, and that was to cross above Falmouth, making use of the fords above that village and those west of the junction of the Rapidan with the Rappahannock. This line presented many advantages, not the least of which was that the army moving thereon would interpose between Lee and the Union capital. The river-banks are densely timbered to the water's edge, and the gentle, undulating country on the south side offered no special advantage to the defending army. Behind the screen of forest it would be possible to move the whole invading force free from observation; and in any case, as the nearest ford is only seven miles distant by road from Falmouth, to gain a long start on the Confederate right wing at the Hop Yard and Port Royal.

This ford once seized, and a strong body of troops thrown across, a passage would be secured for the remainder of the army, and the pontoon bridges might be laid without molestation. Above all, the movement would effectually turn the strong Fredericksburg position, and an opportunity would be presented of crushing the enemy's left wing in detail; or, if Lee preferred to retire rapidly, and to concentrate further south, of bringing him to battle on a fair field.[11]

In May, 1863, Hooker, under exactly similar circumstances, partially adopted this line, and his complete success thereon up to the moment of collision at Chancellorsville proves conclusively that Burnside ought to have anticipated him.

11. The river was unfordable below Falmouth, and one division, therefore, would have been sufficient to protect the line of communication with Aquia Creek against counter-attack.

The demonstration at Skinker's Neck had induced Lee to post a large portion of his force in that distant locality; there were no troops save cavalry and horse-artillery near the fords, and with ordinary precautions the passage might have been easily effected.

The overwhelming objection to crossing at Fredericksburg was that if Lee were warned in time to bring in his outlying divisions to Longstreet's assistance, the Federals would be compelled to assail an exceedingly strong position held by the united strength of the Confederate army. The chance of a successful surprise was very slender; there can be no question, therefore, as to which course Burnside ought to have adopted. Everything pointed to it, and it is almost incredible that he should have neglected so obvious a solution of the problem which confronted him, *i.e.* to cross the river without loss, and to engage the enemy with a good prospect of success.

Unfortunately for his reputation, however, he had too readily assumed that his adversary, alarmed by the attempt at Skinker's Neck, had posted a large force on that flank, and that, expecting attack to come from that quarter, he would be slow to remove it. The wish was father to the thought. He was at no pains to verify the truth of his assumption, and of the presence of two strong divisions near Guiney's Station,[12] as well as of the actual strength of either wing, he remained in ignorance. However, he was perfectly aware that his adversary's army was disseminated; that one portion held the Fredericksburg position, and that another was from twelve to eighteen miles distant.

The Fredericksburg force was within easy reach, and according to his assumption was unsupported; it appeared possible, therefore, to cross the river at that point, and to crush the Confederate left before the right could come up.

The opportunity was tempting, so much so that a more experienced soldier would have suspected a snare, and have divined that the task, seemingly so easy, was exactly what the enemy most wished him to attempt. Burnside, however, wofully underrating the genius of the Virginian general, and believing him completely outwitted, resolved to take advantage of the apparently rash dispersion of Lee's divisions, to cross rapidly at Fredericksburg, and defeat the force there posted.

12. A. P. Hill, 11,000 strong; Taliaferro, 5,000 strong.

For the second time he acted upon a pure assumption. He appears to have confidently expected that if he surprised the passage of the river, the Fredericksburg force would await his onset, and give him the opportunity of beating it in detail. It was within the bounds of possibility that he might cross so swiftly and suddenly as to interpose his army between the widely divided Confederate wings, but what conceivable inducement had he to think that if such a movement seemed likely to succeed the left wing would await attack? Of what paramount importance were the town of Fredericksburg and the line of the Rappahannock to the South that Lee would endeavour to hold them at any cost? If his whole army were concentrated along the formidable Fredericksburg position, the Confederate general would be ready enough to hold his ground and there join issue; but if the Federal advance were so rapid as to prevent concentration, was it likely that he would allow one single Virginian rifleman to throw away his life in useless resistance? The Confederates drew but scant supplies from the ravaged and exhausted district, and Lee would in any case prefer to surrender some miles of territory than to jeopardize the safety of that army on which depended the existence, not only of the Southern Capital, but of the Republic itself.

It may be urged that if Burnside had succeeded in surprising a passage at this point, and the enemy's left wing been compelled to retire, he would have attained one end at least—that of establishing himself on the south bank of the river. Perfectly true; but he ran great risks in the attempt, which by crossing at the fords above Falmouth he would have altogether avoided.

To summarize Burnside's faults: firstly, he underrated his antagonist; secondly, he neglected to reconnoitre as far as was within his power; thirdly, in preference to a line of operations which was feasible and safe, he selected one which promised no more certain result, and which might possibly lead to terrible disaster.

Not once but often, it is true, has the rule that forbids attack with a river in the rear been disregarded by the great masters of war, but seldom without fair prospect of success. If Lee had time to concentrate at Fredericksburg—and even Burnside must have admitted it was quite probable that he would—what hope was there the Confederate general with 78,000 men, would succumb to 110,000 Federals,[13] when at the Antietam, but three short months previous, with an exhausted army of less than 40,000, holding a less favourable position, he had repulsed the assault of more than twice his numbers?[14] Bearing in mind that there was an alternative course open to Burnside, and that he was not even bound to advance at all, is there another instance in history where a general, free to act, ran so great a risk with so little justification?

It is scarcely necessary to discuss at length the further development of his plan, and the course he intended to follow after gaining possession of the heights. Between that point and Richmond lay sixty miles of river, hill, and forest, a wilderness without towns or highways, and abounding in naturally strong defensive positions. The season was adverse, the population hostile, and the roads in bad order.

Little subsistence was to be obtained from the country beyond the river; the district, always thinly populated and sparsely cultivated, had become exhausted and impoverished. He had therefore ordered a huge wagon-train to be prepared with supplies sufficient for twelve days. It was absolutely necessary to be able to move independently if he were compelled to leave the railroad, or if the line were cut or menaced by the Confederates; but at this season, with the roads liable to be made almost impassable by the breaking-up of the frost, it would have been extremely hazardous to quit the regular line of supply.

The reader can estimate for himself the chances of success. He will probably come to the conclusion that Burnside lacked not only the ability, but also the time, the men, and the sup-

13. These numbers give the total, not the effective, strength of either army. A percentage must be deducted for the non-effectives, composed of men sick, absent, and on detachment. The effective strength of the Confederate army was probably about 68,000; of the Federals, 100,000.

14. It must be remembered, too, that the Federal leader believed Lee's strength to be very little inferior to his own. ["Burnside's dispatches contained none of the exaggerated estimates of the enemy's strength which had been so common with McClellan. . . . In fact, he said nothing whatever about the size of Lee's army, which he probably put at about its actual figure of some 85,000" (Kenneth P. Williams, *Lincoln Finds a General: A Military Study of the Civil War* [New York, 1949], II, 518).]

plies, necessary to achieve his purpose, and that the Confederate Capital was in little danger.

During the war no less than six Federal generals attempted to reach Richmond through Northern Virginia. McDowell, Pope, Burnside, Hooker, Meade, and Grant recoiled successively, as did the marshals of France before Wellington, from the invulnerable front of the Confederate army. In but one campaign did the Confederate numbers exceed those of the Federals, and then victory was complete; in the rest they were far inferior. Can further proof be needed of the genius of Lee and the stern valour of his soldiers?

Nor was it brilliant strategy or the irresistible onset of disciplined valour that at last gave the victory to the North. Not the public foe but her own exhaustion wrought the fall of the Confederacy, and the utmost limit of endurance had been reached on that day when, after four years of battle, the Army of Northern Virginia laid down its arms, "non victus sed vincendo fatigatus."

To return to Burnside. Having determined to cross the river and carry the heights beyond Fredericksburg before Lee could concentrate thereon, he had before him an operation which rendered necessary great precaution and much nice calculation. Time was the very essence of the contract; success was altogether dependent upon rapidity of movement.

Part of the Confederate army was several miles distant,[15] and the country roads were in bad condition. Fifteen hours at the very least must therefore elapse after the Federal purpose was exposed before the whole force could come together above Fredericksburg, and ten hours before Early's leading battalion could arrive from the Hop Yard.[16]

To bridge the Rappahannock, 200 yards wide, and within easy range of the enemy's guns, even if the hostile picquets were driven in, and the operation covered by artillery posted on the

15. Early's division, 7,500 strong, near the Hop Yard, twelve miles from Fredericksburg; D. H. Hill's division, 9,000 strong, near Port Royal, eighteen miles from Fredericksburg.

16. The rate of marching on the bad roads would not exceed two miles an hour. It would take two hours to convey the order from headquarters, and two more to call in outposts and forage parties, strike camp, feed, load wagons, and fall in. More than one road was available for the march of either division.

Stafford heights, was not to be done in a moment. Burnside proposed to lay six bridges, three under cover of the town, and three below the mouth of Deep Run; and it was necessary that these should be firmly established before he could venture to attack.

By renewing the demonstration near Skinker's Neck it was possible to deceive Lee as to the real point of passage, and to retard, for some time at least, the withdrawal of his troops from that neighbourhood (Burnside did not fail to make use of this auxiliary operation; but the feint did not mislead his sagacious adversary, and no Confederate reinforcements were sent from left or centre to the threatened point.) The river spanned, several hours would be consumed in the transit of the troops, and in placing them in line of battle. It must have been quite clear, therefore, to the Federal staff that, even if all went well, they would have but a very narrow margin of time to spare. A few hours' delay, the design once exposed, would enable Lee to bring up his right wing into line; in which case they would have a choice of difficulties, either to attack a strong position held by the whole Confederate army, or to retire across the river in the presence of a skilful and determined enemy.

The object of the Confederate leader's strategy and the disposition of his troops have already been examined (pp. 20–21). He had evidently considered and prepared for every eventuality, and was ready to concentrate at any point between Fredericksburg and Port Royal, or in the interior of Spotsylvania County, as the enemy's movements might dictate.

Why he detained D. H. Hill at Port Royal does not appear. His sources of information were excellent, and he was doubtless well aware that Burnside would not attempt to operate below the Hop Yard. The natural course would have been to withdraw this strong division from the distant point, and to place it within the limits of possible operations, say near Guiney's Station. [Map II.] Whether the position was maintained only for convenience of supply, or whether with the deliberate purpose of enticing Burnside into the toils at Fredericksburg, there is, unfortunately, no direct evidence to show.[17]

17. [Freeman seems to have been of the opinion that Lee wanted to entice Burnside into attacking him. "The position directly at Fredericksburg was so strong that elabo-

Topography of the right bank of the Rappahannock.

It is now necessary to describe particularly the topography of that bank of the Rappahannock to which the Federal commander proposed to transfer his force, and of that position on which his opponent hoped rather than expected to meet his onset. [Map III.]

The peculiar terrace formation of the terrain is a feature familiar to geologists, being generally observable along the course of great rivers. There are four distinct elevations, indicating, probably, the successive levels of the Rappahannock. The first stage is the narrow strand left by the stream receding as its channel deepens. The second, thirty feet above, is a broad alluvial plain, the region of cultivation. The third is a low and narrow ridge which touches the river opposite Falmouth, curves gradually away to an extreme distance of 4,500 yards, and approaches it again beyond the Massaponax, enclosing in its chord the fertile area of the lower level. Half-a-mile in rear rises a loftier and thickly wooded crest, terminating an extensive table-land; and between these two ridges are numerous deep ravines, carved out by the action of the mountain streams. The left bank of the Rappahannock, held by the Federals, is crowned by a bold line of bluffs, 150 feet above the level of the stream, and beyond, stretching away towards the Potomac, lies the broad plateau of Stafford County. Along the lower ridge of the opposite bank, immediately overlooking Fredericksburg and the surrounding plain, was the main position of the Confederate army.

The five principal spurs of this ridge are:—

> Taylor's Hill, above the river and facing Falmouth; 50 feet high.
> Stansbury Hill, 1200 yards S.E.; 50 feet.
> Marye's Hill, covering the main issues from the town; 40 feet.
> Lee's Hill, 900 yards to the right rear; 90 feet.
> Prospect Hill, overlooking the Massaponax; 40 feet.

Behind Fredericksburg the surface of the ridge is barren and treeless; the slopes are smooth and grassy, and the gradients

---

rate fortifications would have convinced any antagonist that it was impregnable. . . . It doubtless was better, in Lee's opinion, to invite attack by seeming negligence, than to discourage it by a show of complete preparation" (Douglas Southall Freeman, *R. E. Lee* [New York, 1935], II, 441–42).]

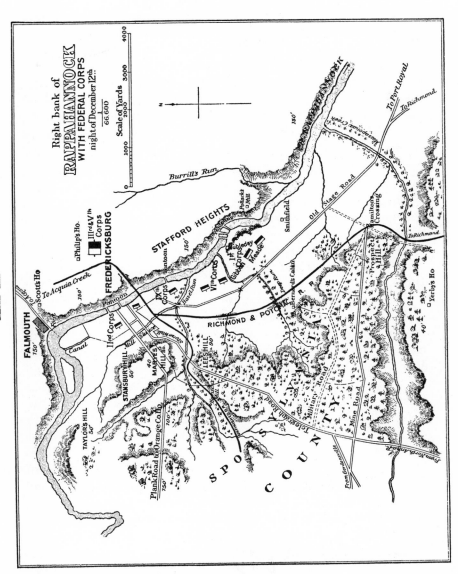

generally easy; from Lee's Hill to the Massaponax the heights are covered with a thick growth of pine and oak, which, except at one point to be hereafter mentioned, does not anywhere encroach upon the plain.

Along the front of the position, from the Rappahannock to the Hazel Creek, a mill-sluice, twenty feet wide and four deep, runs midway between the upland and the town. Opposite Stansbury Hill, the depression through which it flows is shallow, and the water-way becomes a formidable obstacle; but opposite Marye's Hill the hollow is of such breadth and depth as to shelter advancing troops from direct fire, not only when crossing the two bridges by which it is spanned, but also for some distance beyond.

Marye's Hill is a steep and abrupt salient, on which a conspicuous object is the picturesque mansion of the Marye family. The ground between the foot of the hill and the mill-sluice, a space of 400 yards, is flat and open, and was broken only by some plank fences and a few wooden houses dotted here and there.

The plain extending from Hazel Run to the Massaponax is of the same character, encumbered by a few buildings, but too soft and cut up by too many creeks and ditches to admit of cavalry movements upon its level surface. Hazel and Deep Runs flow through ravines quite thirty feet in depth, and hid by timber and dense undergrowth. The Massaponax Creek, a stream of larger volume, runs swiftly between wooded and marshy banks, impracticable for troops. The hills to the south are forest-clad, and correspond in height with those fringing the opposite side of the Massaponax valley, the breadth of which, at the entrance, is 500 yards. Above Falmouth, some miles in rear of Stansbury Hill, the Rappahannock may be crossed by two easy fords,[18] but between that village and Fredericksburg there exists one only, and that difficult and dangerous.

It will be observed that the Stafford plateau so completely commands the plain and lower ridge on the opposite shore, that it would be difficult and costly for a force acting on the right bank to effectually prevent an enemy, if he were to cover the operations by strong batteries posted along this dominating

18. See Map II.

crest, from laying bridges and establishing himself below the town.

Fredericksburg is a small and picturesque city, containing nearly 5,000 inhabitants. It possesses many substantial brick houses and public edifices, but the greater part of the buildings are of wood. In appearance and condition it has little altered since the war, nor indeed since the old colonial days. Three principal roads issue from it, and two railway tracks:—

1. The Orange plank road passes through the outlying houses and corn-fields between Marye's Hill and the town, and leads westward over the opposite heights.

2. Two hundred and forty yards south the Telegraph road keeps a parallel direction for some 600 yards; meeting the steep front of Marye's Hill, it turns abruptly to the left, skirting for nearly half a mile the base of the salient; inclining again to the left it crosses Hazel Run, ascends the slope of Lee's Hill, and bearing S.S.E. continues its course behind the crest of the lower ridge.

3. The Old Stage road to Richmond makes its way south, between the river and the railroad, to a point below the Massaponax, where it divides, branching towards Richmond and Port Royal. This highway passes between earthen banks, three or four feet high, stiffened, or as it were revetted, by the tough roots of the cedars which line it.

Never in good order, these roads were much cut up by the heavy traffic of the war. They are about twenty-five feet wide, and the soil is clayey.

4. Leaving the town at the S.E. angle, the Richmond and Potomac railroad turns in a southerly direction, and running for some miles nearly parallel to the river cuts the plain in two. The track passes over an embankment from three to four feet high.

5. A second embankment, 550 yards long, carried an unfinished single-gauge line over the depression of Hazel Run.

The stone bridge, by which the Richmond and Potomac railway spanned the Rappahannock, had been destroyed early in the war, and no other bridge existed.

We will now consider the means taken by the Confederates to improve the natural advantages of their position. As early as November 23rd, twenty-four hours after the arrival of Longstreet's Corps, Lee had ordered the work to be taken in hand. The first thing to be done was to secure the guns from the fire of the enemy's pieces on the Stafford Heights. Under the superintendence of Brigadier Pendleton (a minister of the Episcopal Church), Chief of Artillery, forty gun-pits were constructed along the crest of the grassy upland between the river and Lee's Hill. The work progressed but slowly, for the ground was hard frozen, and tools were scarce.

It was necessary that the approaches from Fredericksburg, the Telegraph and Orange plank roads, should be effectually blocked: with this view nine gun-pits were sunk on Marye's Hill, along a front of four hundred yards, and four beyond the left shoulder of the spur so as to enfilade the plank road as far as the town. Several pits were dug on Stansbury Hill and the adjoining high ground, and a small work was thrown up between the foot of the slope and the mill-sluice from which a flank fire might be brought to bear on a force advancing from the town by either roads.

Cover for infantry was provided as follows:—

The Telegraph road, where it skirts the base of Marye's Hill, is hollow, and is sustained on either side by a stout stone wall. A trench was dug on the road-side nearest the town, the earth thrown over and banked up against the outer fence. This wall is rather more than half a mile long, and about forty feet below the crest of the hill, which rises steeply and abruptly above it. The house on the plateau is built of brick, and there are several stone fences enclosing the premises. [Sketch I.]

Between the Orange and the Telegraph roads, in prolongation of the stone wall, a shelter trench was dug, which is still to be seen; it was doubtless improved after the battle of December 13th. [Sketch I.]

Half a mile in rear of Marye's Hill, on the slope of the further ridge, a second line of earthen parapet gave cover for a brigade, and enfiladed the Orange road.

At the foot of Lee's Hill and astride the Hazel Run log breastworks were raised; and this section of the position, as well as the central re-entrant, was also protected by abattis.

No further artificial cover existed. It has been stated by numerous writers that at the first battle of Fredericksburg the whole Confederate front bristled with entrenchments, but in contemporary records no confirmation of this assertion is to be found, and Burnside himself, in his evidence before a Committee of Congress, testified that the defences were but slight. The elaborate series of works which afterwards lined the hills was constructed after the engagement. Not that the right section of the position, from Lee's Hill to the Massaponax, was neg-

SKETCH I

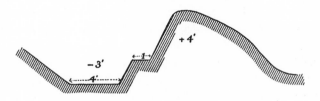

INFANTRY SHELTER TRENCH.

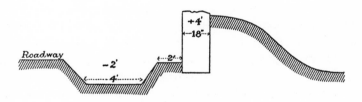

STONE-WALL PREPARED FOR DEFENCE.

lected. It will be remembered that there the whole surface of the hills is covered with a dense growth of timber, and Lee thought it more essential to employ his men in cutting roads through the woods, and in laying a line of field telegraph, than in constructing breastworks or redoubts. The task of creating good lateral communication was the most important. Cover to a certain extent already existed, and the ground in front was clear and open; good roads in rear of the crest of the plateau were the one thing necessary to enable the general-in-chief, by rapidly transferring along them reinforcements and reserves, to

The Campaign of Fredericksburg 49

effect the primary object of all combinations and of every system of defence, *i.e.*, to meet the enemy with superior numbers at the desired point. Existing tracks were improved, gaps opened in banks, fences thrown down, and ditches filled, in order to facilitate the passage of infantry and artillery.

The houses and buildings in Fredericksburg overlooking the river were loopholed. Rifle-pits were constructed on the edge of the bluffs; and along the Old Stage road, below the point where it is joined by the track from Hamilton's Crossing, shelter-trenches were dug by Stuart's troopers. The embankment of the Richmond and Potomac railway forms a stout breastwork, and was so used by the skirmish line of the right wing in the ensuing battle.

To prevent gunboats ascending the river, a battery was placed in an earthwork four miles below the town.

The weak points of the position were:—

1. It was commanded—except the central re-entrant—by the opposite plateau.

2. This command, and the natural entrenchment of the stage road, would make it a hazardous undertaking to assume the offensive. An enemy, holding the Stafford Heights and the road, would retire, if repulsed, as it were to a citadel.

3. The ground was soft and intersected by water-courses, and did not therefore admit of cavalry action.

4. The shortest and easiest line of supply and communication with Richmond, the Richmond and Potomac railway, was in prolongation of the right flank; and the main line of retreat, the Telegraph road, ran in rear of that flank and not of the centre.

5. It could be turned by the fords above Falmouth.

6. As the river interposed between the armies, and as the left bank was wooded in parts and higher than the right, the enemy could move in either direction on his own side of the stream without being observed.

7. The bend of the river above Falmouth made it possible for guns posted on the heights near that village to enfilade the lower ridge as far as Marye's Hill.

The strong points:—

1. The length of the line selected was well proportioned to

the numbers available for defence, allowing 11,000 men to the mile, or rather less than six to the yard.[19]

2. Both the lower and further ridges afforded many good artillery positions, commanding the length and breadth of the plain below.

3. The ground between the hills and the river was open and unobstructed, a fair field both of view and fire.

4. The ravines between the ridges gave good shelter, close at hand, for supports, reserves, field hospitals, and ammunition carts.

5. The thick screen of forest and the depth of the ravines served to conceal not only the movements, but also the numbers, of the defending troops and made it practicable to mass in unexpected strength at any given point.

6. The lateral communications were good, and had been skilfully improved and supplemented.

7. Communication with Richmond was direct and rapid, and the supply depôts on the Virginia Central railway were within easy reach, Gordonsville, the principal magazine, being forty-six miles distant by road.

8. There was a strong second line, nearly three-quarters of a mile in rear, along the crest of a loftier plateau, beyond the range of artillery on the Stafford Heights.

9. In case of retreat the country abounded in strong rear-guard positions.

10. The flanks rested on strong natural obstacles.

11. While the fords were securely held, and the river-banks vigilantly patrolled, it was scarcely possible for the enemy's cavalry to cross in force for purposes of reconnaisance or raid.

12. The enemy's force once transferred to the right bank, his after-movements would be plainly visible.

13. Artillery posted on the lower ridge would render the operation of crossing hazardous and costly.

14. The several bastion-like spurs gave reciprocal flank defence.

15. The enemy would have to fight with a deep and broad river in his rear.

[19]. "Including all arms and reserves, it may be assumed, as a general rule, that the force for the defence of a position should be equal to five men for each yard in extent." *Field Exercise*, 1884.

16. About Fredericksburg an attacking force would be compelled to deploy under artillery fire at short range, and its advance would be seriously obstructed by the mill-sluice.

17. The various creeks which intersect the ground in front of the centre and right, over which the Federals were almost bound to advance, would impede communications and the exact execution of combined movements.

Having now examined Lee's position from his own point of view, let us cross the Rappahannock and endeavour to realize the prospect that lay before Burnside, and the deductions he ought to have drawn therefrom.

Looking west from the windows of the Philip's House, a planter's mansion on the Stafford Heights, distant about 3,000 yards as the crow flies from the crest of Marye's Hill, where he had established his headquarters, he could command at single *coup d'œil* the whole panorama of the field just described. Before him lay an open plain, narrow to the north, expanding to a greater breadth to the south, and dominated by a double tier of hills.

Along the bare surface of the lower ridge between Hazel Run and the river, earthworks were plainly visible, and below to the left the highlands were covered with dense woods, which his observation could not penetrate. Upon the grassy uplands beyond Fredericksburg, working parties were busily turning up the yellow clay; sentries and picquets watched the river-bank, and across the plain gray-coated vedettes passed to and fro: no large body of men nor encampment was visible, yet who could say how large a force might not be screened by that dark line of forest, or concealed in the deep depressions of the hills? He had little information as to the number or dispositions of the enemy, and it must have been obvious that before he could advance against the heights confronting him, it would be necessary to pass over the river every available man, for there was always a chance, however sudden and rapid his movements, that Lee would be enabled to concentrate against him his whole force.

The Federal sentries upon the edge of the Stafford plateau looked down into the streets of the little town; and there, if heavy batteries were brought forward to cover the operation, it would be no hazardous or prolonged task to bridge the stream. But so contracted is the space between the river and the ridge, and so exposed was it in every quarter to the fire of the Confederate guns, that the deployment of the attacking force would be difficult and costly; and, unless those guns were disabled or driven from the hills, the town itself would prove a mere "shell-trap" to the troops in possession. Away to the left, however, where the highlands stand back at a distance of 4,500 yards from the river bank, and the wider strand below the cliffs forms a spacious landing-place, there was room to bridge, to cross, and to manoeuvre unmolested; there, too, the Old Stage road, with its double embankment, presented a strong *place d'armes* and base of attack.[20] Moreover the enemy's right, though *appuyed* on a difficult obstacle, was his weak flank, strategically and tactically, nearest his base and covering his communications, undefended by entrenchments, unprotected by obstruction in front. It was at this flank, if the Federal general desired a decisive victory, that he must strike his fiercest blow. Could he make good his footing on the slopes, within the dense woods he would find himself at once, as regards position, upon equal terms with his antagonist, and, if superior in numbers, might await the issue with confidence.

Preliminary operations.

On December 10th, Burnside determined to cross the river and attack the Confederate position; moved thereto, say his apologists, rather by the outcry of the Northern press, clamouring for instant action, than by the approval of his own judgment. Yet it is certain that he still cherished the delusion that Lee was over-reached, and had exposed his army to be dealt with in detail.

On that date the infantry moved into position directly behind the crest of the plateau above Fredericksburg; Sumner on the left, Hooker in the centre, and Franklin opposite Deep Run.[21]

20. [*Place d'armes*—literally parade or drill ground; in this instance, "An open spot on which troops can form in proper order as the successive bodies arrive" (Theodore Ayrault Dodge, *A Bird's Eye View of Our Civil War* [new and rev. ed.; Boston, 1897], p. 337).]

21. At the same time a small force stationed near Skinker's Neck was ordered to renew the demonstration of crossing at that point, and troops were marched along the open Stafford plateau as if en route to reinforce it in strength. [Sumner's Right Grand Division occupied the *right*.]

The pontoons were lowered to the water's edge. Three days' rations were issued to the whole army; the infantry carried sixty rounds of ammunition per man; and the huge wagon-train, with supplies sufficient for twelve days, was held in readiness to follow the advance. The wagons of a division, states an officer of the Federal Commissariat Service, covered three miles. There were eighteen infantry and cavalry divisions in Burnside's army. The train, therefore, if moving along one road, would have been nearly sixty miles in length.

To cover the passage, 143 guns were posted behind epaulements along the crest of the Stafford Heights; distributed as follows:—

From beyond the road leading through Falmouth to a point 500 yards below (including 6 twenty-pounders)..................... 40 guns.
From this point to opposite the centre of town.................. 36 "
On the ridge south (including 7 four-and-a-half inch siege guns)... 27 "
To Pollock's Mill... 40 "

Three bridges were to be thrown across the stream under shelter of the town; three about a mile below, near the mouth of Deep Run.

Before drawn on the 11th, the pontoniers began the work. At three o'clock the report of two signal guns announced to the Confederate army that hostile movements had begun, and Longstreet's troops, breaking up from their bivouacs, marched to the positions already allotted to them, viz.:—

Anderson's division, from Taylor's Hill to the Orange road.

Ransom's and McLaws's, Marye's and Lee's Hills and the Hazel Run defile.

Pickett's and Hood's, the wooded slopes as far as Hamilton's Crossing, a little wayside station on the Richmond and Potomac line.

Barksdale's Mississippi brigade[22] of McLaws's division was posted within and on either side of Fredericksburg; an advanced line of two companies and several strong detached parties being pushed forward to the verge of the bluffs.

No message of recall was sent to Jackson.

A heavy fog hung upon the water, and opposite the town the Federals worked for some time unmolested. Not until the

22. About 1,600 strong.

bridge was two-thirds completed, and shadowy figures became visible in the mist did the Confederates open fire. At such close quarters the effect was immediate, and the builders fled. Twice at intervals of half an hour they ventured again upon the deserted bridge, twice were they driven back. After the third repulse Barksdale ordered up the remainder of his brigade, and prolonged his line to right and left.

Strong detachments were now moved forward by the Federals to cover the working parties, and artillery began to play upon the town. Securely posted in rifle-pits and cellars, the Mississippians were not to be dislodged; but the covering troops, exposed on the bare slopes of the cliff, lost heavily, nor could the working parties live upon the bridge.

At ten o'clock Burnside ordered the heavy batteries to concentrate their fire, and every gun that could be brought to bear on Fredericksburg discharged fifty rounds of shot and shell. This bombardment, to which the Confederate artillery did not reply, lasted upwards of an hour. Though the effect on the buildings was appalling, though flames broke out in many places and the streets were furrowed with round shot, the defenders not only suffered but little loss, but at the very height of the cannonade easily repelled another attempt to complete the bridge.

After a delay of several hours, Hooker, commanding the Centre Grand Division, recognizing that so long as the enemy held the town, the passage could not be achieved, called for volunteers to cross the stream in boats, and to drive the enemy from their cover at the point of the bayonet. Four regiments of Getty's division responded to the summons.[23] A portion descended the cliff, and, from behind what shelter they could find, for fifteen or twenty minutes assailed the defences of the Confederate sharpshooters with a brisk fire. Ten pontoon boats were then manned, and though many lives were lost during the transit, the gallant Federals pushed quickly across and won

23. [Henderson is in error in crediting Hooker with sending volunteers across the Rappahannock in boats. Burnside himself gave the order, and four regiments of Sumner's Right Grand Division responded: the 89th New York Volunteers of Getty's division, and the 7th Michigan, the 19th and the 20th Massachusetts of Howard's division (*Official Records*, XXI, 265, 282, 331).]

the shelter of the bluffs. Other boats followed; and Barksdale's brigade, which had no orders to hold the place against an advance in force, retreated skirmishing to the upper town. So rapid, however, were the enemy's movements, and so defective the means of communication between the buildings occupied, that one hundred Mississippians were cut off and captured. At about 4:30 p.m., three bridges being at last established, the Federals pushed forward, and the Confederates, retiring in good order, evacuated the town.

Franklin, a mile below, thanks to the configuration of the river bank, and to a diversion he had made lower down the stream, quickly drove the sharpshooters from the rifle-pits, and by one o'clock had constructed three bridges. As Hooker and Sumner were unable to act in concert with him, he passed over only a small part of his force during the afternoon.

We have already observed that Burnside's only chance of success lay in prompt action. The rivet which held his plan together, and without which it would utterly fall to pieces, was celerity of execution. Barksdale's stubborn resistance had already caused much delay, and it was doubtful if the concentration of Lee's army could now be anticipated. We should expect to hear that, immediately the bridges were completed, he at once set about transferring his troops across the stream, and placing them in readiness to move forward with the earliest light; and doubtless the Southern sentries listened from sundown to dawn for the tramp of countless feet, the rolling of heavy wheels upon the frozen roads, for the deep murmur and "the stifled hum" which betray the march of an advancing host. But what was the case? The long hours of darkness slipped peacefully away, no unusual sound broke the silence of the night, and all was still along the Rappahannock. Had Burnside then abandoned his attempt? Had he, recognizing that no good result could come of the movement, resolved at the last moment to hold to his own opinion, to defy the clamour of the press, and not to hazard, at such a call, the existence of his gallant soldiers? Unfortunately, no such praiseworthy determination inspired him: as has been said of another, "his generalship arrayed every chance against him."

Although every moment of delay saw hope of success grow

fainter, there was no question of relinquishing the enterprise. No movement took place during the night, but on the next day, December 12th, all Franklin's and the greater part of Sumner's Grand Divisions, covered by a dense fog, leisurely crossed the river and took up their stations on the right bank.

Sumner occupied Fredericksburg and the open space between the town and Hazel Run; Franklin held the line of the Old Stage road, along which entrenchments were rapidly thrown up, as far as the steep banks of the water-course which bears the name of Deep Run. A cavalry reconnaissance, pushed forward as far as the railroad, came into collision there with Hood's outposts and Stuart's scouts, but the Federal horsemen were driven back.

As soon as Lee, on the morning of the 12th, found that the bridges were completed, he recognized, from their very position, that his adversary, bound as he was to these two narrow lines of retreat, could not extend his line so far as to operate beyond the Massaponax without unduly weakening his centre. The indication was clear that the battle would be confined to the ground north of that creek; A. P. Hill and Taliaferro were therefore called up from Yerby's and Guiney's Station respectively, relieving Hood and Pickett about noon, and occupying the space between Hamilton's Crossing and Deep Run. Hood and Pickett, closing on the centre, filled the interval between Deep and Hazel Runs.

The gallant defence of Fredericksburg had frustrated the intention of the Federal general to surprise a crossing, and to attack Longstreet before Jackson could be brought up; and Lee, with ample time at his disposal and confident in the wisdom of his dispositions, calmly awaited the development of his adversary's plans. Not until noon on the 12th, satisfied at last that the whole of the Federal army was crossing before him, and that the operation in progress was no feint, did he despatch couriers to call in his distant divisions.

Shortly after dawn[24] on the 13th, D. H. Hill and Early ar-

24. About eighteen hours after the order had been issued, but the divisions did not begin their march until after nightfall. [A. P. Hill stated in his report that he was ordered "on the evening of December 11 . . . to move my division at dawn on the 12th" (*ibid.*, p. 645). Taliaferro's division marched "on the morning of the 12th" (*ibid.*, p. 675).]

rived from Port Royal and the Hop Yard, after a night march of eighteen and twelve miles respectively. These divisions were stationed in rear of the Light Division of A. P. Hill, and the whole Confederate army was now concentrated along the ridge between the Rappahannock and the Massaponax.

It is possible that the summons to Hill and Early was not issued until noon, that they were suffered to remain *en évidence* along the river-bank, and that their march was delayed till night, in order to foster Burnside's delusion, to prevent him taking the alarm and retiring across the stream.

While Franklin and Sumner were crossing on the 12th, the Federal batteries maintained a heavy fire, but no reply was elicited from the Southern artillery. Lee, it has been remarked, feared that even at this, the eleventh hour, the foe, if roughly handled, might withdraw from the snare he had so incautiously approached; and was desirous, moreover, that the position of his guns might remain unknown until it was absolutely necessary to reveal them.

The reason of Burnside's untoward delay has never been satisfactorily explained. It may be imagined that he and his staff were incompetent to manoeuvre rapidly so vast a force, or that the force itself was ill-trained and unwieldy, but the after-events completely dispose of these suppositions. It has been asserted that the fog which hung over the river on the 12th prevented his advance, but this statement does not account for his quiescence on the night previous. Fog or no fog, it was clearly his duty to have had his men in position for attack early on the 12th. However it may be, it is certain that much precious time was cut to waste, and that his adversary received ample warning of his intent.

It may be that want of definite information as to the numbers, disposition, and movements of the opposing force was the cause of his inaction on the night of the 11th, and of his determination to attack on the 13th, when all hope of surprise had passed away. On the 10th of December, he had learnt from an escaped slave that a road ran in rear of the crest of the opposite ridge, and on the 11th, from a German prisoner, that the commanding heights beyond were partially occupied as second line of defence; but beyond this he knew little. In contradistinction to the enterprise of Stuart's brigadiers, no attempt was made, even by small scouting parties, to penetrate the hostile lines. The Confederate troopers on either flank had ridden far and wide over the Virginian counties in rear of the Union outposts. At the end of November, Hampton had captured two whole squadrons; on the 11th of December, had cut off a convoy of wagons near Dumfries, on the Potomac, fifteen miles above Aquia Creek;[25] and prisoners had been taken at various points within the Federal lines. It is true that Burnside, from his commanding position on the Stafford Heights, and with the aid of balloons, was able to form a fairly accurate idea of the strength of the Fredericksburg position, and also of the whereabouts of portions of Lee's army; but a few determined men, working on the far side of the river, would have supplied him with information of the utmost value. From their reports he would probably have discovered that the position—naturally formidable— was thoroughly prepared for defence, that one-half of Jackson's corps was in close support of Longstreet, and that the demonstration at Skinker's Neck had not prevailed on his adversary to weaken his left by sending reinforcements to that quarter.

THE BATTLE OF FREDERICKSBURG.

Like its predecessor, the 13th of December broke dull and calm, and the mist which shrouded the river and the plain hid from each other the rival hosts. [Map IV.] Long before daybreak the Federal divisions still beyond the stream began to cross; and as the morning wore on, and Franklin's line moved forward from the bivouacs, the rumbling of artillery, the loud words of command, and the sound of martial music, came muffled by the fog to the ears of the Confederates lying expectant on the heights. Now and again the curtain lifted for a moment, and the Southern guns assailed the long dark columns of the foe. Very early had the Confederate army taken up its final position. On the extreme right, 200 yards in rear of Hamilton's Crossing, fourteen guns, under Colonel Walker, were stationed on the spur called Prospect Hill. Supported by two regiments of Field's brigade, these pieces were held back for the present within the thickets.

25. [The action at Dumfries, Virginia, took place December 12 (*ibid.*, pp. 689–91).]

The massive foundation of the railroad, though it formed a tempting breastwork, was only utilized as such by the skirmishers of the defence.[26] The border of the wood, 150 to 200 yards in rear, looked down upon an open and gentle slope, and along the brow of this natural glacis, covered by the thickest timber, Jackson posted the main body of his fighting line. To that position it was easy to move supports, unperceived and unopposed; and if the assailants were to seize the embankment,

26. The skirmish line of Jackson's corps is not shown in the map.

MAP IV

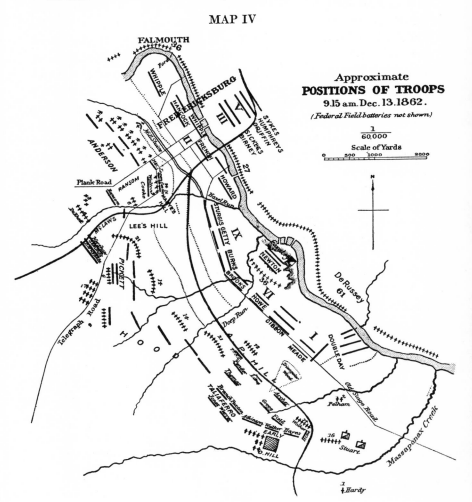

he relied upon the deady rifles of his infantry to bar their further advance up the bare ascent beyond.

The Light Division, under Ambrose Hill, formed the first line of Jackson's corps. To the left of Walker's batteries, posted in a trench within the skirt of the wood, was Archer's brigade of seven regiments, including two of Field's, the left resting on a coppice that projected beyond the general line of forest. On the further side of this coppice, but nearer the embankment, lay Lane's brigade, an unoccupied space of 600 yards intervening between his right and Archer's left. Between Lane's right and the edge of the coppice there was a stretch of open field 200 yards in breadth. Both of these brigades had a strong skirmish line pushed forward along and beyond the railroad. Five hundred yards in rear, along a military road, Gregg's brigade covered the interval between Archer and Lane. On Lane's left rear was Pender's brigade, supporting twelve guns posted in the open, beyond the embankment, and twenty-one massed in a field to the north of Bernard's cabin. One regiment (the 7th North Carolina), drawn up along the railroad, 200 yards in rear of the guns, acted as immediate escort of the foremost batteries. Four hundred yards in rear of Lane's left and Pender's right, Thomas's brigade of four regiments was stationed in support.[27]

The first line of Jackson's Corps was thus held by one division, three brigades in front and two in support; fourteen guns were massed on the left, and thirty-three on the right.[28]

It is necessary to notice particularly the shape, size, and position of the projecting tongue of woodland which broke the continuity of Hill's line, as it influenced greatly the conduct of the ensuing battle. A German officer on Stuart's staff had

27. Approximate strength of brigades:—Field, 1,600; Archer, 2,000; Lane, 2,000; Gregg, 1,800; Pender, 2,000; Thomas, 1,600.

28. [". . . fourteen guns were massed on the *right*, and thirty-three on the *left*." As a matter of fact, no one knows exactly how many guns were placed on the left. According to Jackson, twenty-one guns were posted near the Bernard cabins and twelve additional guns beyond the railroad. His Chief of Artillery, however, claims that only twenty-one guns in all covered the left of Jackson's line (*Official Records*, XXI, 631, 636). The foremost authority on the artillery of the Army of Northern Virginia accepts the latter figure, stating that Henderson followed the errors of previous historians—and presumably Jackson, too—in counting one group twice (Jennings Wise, *The Long Arm of Lee* . . . [Lynchburg, Va., 1915], I, 378 n.).]

the day previous, while riding along the position, remarked its existence and suggested the propriety of razing it; but, though Jackson himself predicted that there would be the scene of the severest fighting, the ground was so marshy within its recesses, and the undergrowth so dense and tangled, that it was judged impenetrable and left intact and unoccupied, an error of judgment which cost the Confederates dear. General Lane had also recognized the danger of leaving so wide a gap between Archer and himself, and had reported to his divisional commander, but without effect.

This salient of the wood was triangular in shape, extending nearly 600 yards beyond the railway embankment. The base, which faced the Federals, was 500 yards in length. Beyond the apex the ground was swampy and covered with scrub, and the ridge, sinking to a level with the plain, afforded no position from which the artillery could command the approach to or issue from this patch of jungle. A space of 700 yards along the front was thus left undefended by direct fire.

His right resting on the railroad at Hamilton's Crossing, and his line extending in a semi-circle behind Archer's and Gregg's, Early had drawn up his division in support of Hill. Three brigades, Hays, Atkinson, and Walker, were deployed in front; Hoke's brigade on the right flank, along the railroad.[29]

Five hundred yards in rear of Gregg, Paxton's, the right brigade of Taliaferro's division, connected with Early's left.[30] Branch's brigade was immediately behind Thomas;[31] Warren and Jones in second line. The division of D. H. Hill, and several batteries formed the reserve, and a portion of Early's artillery was parked about half a mile in rear of his division, in readiness, if necessary, to relieve the guns on Prospect Hill.

Jackson's infantry mustered 30,000 strong, and his line covered 2,600 yards; about eleven men to a yard, including reserve: a deep formation, but this was the weakest flank of the position.

29. Approximate strength of brigades:—Hays, 1,850; Atkinson, 1,850; Hoke, 1,850; Walker, 1,850.
30. Approximate strength of brigades:—Branch [Starke], 1,600; Warren, 1,600; Jones, 1,600.
31. [Starke's brigade, commanded at Fredericksburg by Colonel Edmund Pendleton, and not Branch's brigade, was posted in support of Thomas (*Official Records*, XXI, 686). Henderson possibly confused the two because both officers had been killed at Antietam.]

Opposite Deep Run, on Jackson's left, Hood's division was stationed, and next in order came Pickett's, prolonging the line to Hazel Run. Fourteen guns were massed before each of these divisions, and the level plain in front was covered with skirmishers. This portion of Lee's line, 3,000 yards long, was held by three men to the yard; it was flanked by Jackson's left brigades and the salient of Marye's Hill. Three batteries were held in reserve behind both Hood and Pickett.

Kershaw's brigade of McLaws's division covered the wooded ravine of Hazel Run, with Barksdale's and Semmes's in support. One company was pushed forward along the right bank of the creek, and Howison's mill was held by a battalion. Three regiments of Cobb's brigade occupied the stone wall at the base of Marye's Hill, the 18th Georgia on the right, the 24th Georgia in the centre, and the Phillip's Legion on the left. On the reverse slope, 200 yards behind the crest, Cooke's brigade of Ransom's division was stationed, with the 16th Georgia of Cobb's brigade on the right. In the trench between the Telegraph and Orange plank roads lay the 24th South Carolina of Ransom's brigade of Ransom's division;[32] the remainder of this brigade being entrenched 600 yards in rear, on the slope of the further ridge.[33]

The gun-pits on Marye's Hill were occupied by nine guns, under Colonel Walton, manned by the Louisiana Washington Artillery; in pits to the left of the plank road, was Maurin's battery of four pieces, and six guns were placed in immediate support in the depression behind the right shoulder of Marye's Hill. This important section, covering the roads, was thus held by three brigades in the first line, twelve regiments; three brigades, fifteen regiments, in the second line; in all, 11,000 infantry, and nineteen guns. The actual fighting line was composed of one brigade in front, and three (including Kershaw) in close support; about 7,400 infantry, all of whom took part in the engagement.

Furthermore, on Lee's Hill were twenty-one guns, eight of

32. [The 24th North Carolina Volunteers occupied the trench to the left of the stone wall (*ibid.*, p. 625).]

33. Approximate strength of brigades:—McLaws's division: Kershaw, 2,400; Barksdale, 1,200; Semmes, 2,400; Cobb, 1,600. Ransom's division: Ransom, 1,700; Cooke, 1,700.

which commanded the Telegraph road and enfiladed the embankment of the unfinished railway; and three heavy batteries (two under Rhett) on the slope of the further ridge were ready to sweep the exposed plateau of Marye's Hill, should it be carried by the enemy. In general reserve, seventeen smooth bores were stationed in a gully behind Lee's Hill. The left flank of the position was entrusted to Anderson's division; five batteries in gun-pits on the ridge covered the front, one was held in reserve near Stansbury Hill, and six pieces stood on Taylor's Hill. It was unlikely that this flank would be seriously attacked; it was therefore held by but 7,000 infantry and thirty-four guns, three men to the yard. One battery, occupying the earthwork on the level ground, covered the ford below Falmouth, and at the same time flanked the plank road. Stuart's cavalry and horse artillery, two brigades, 4,000 strong and eighteen guns, on the right flank of the army and well to the front, filled the interval between Hamilton's Crossing and the Massaponax; the skirmish line (dismounted) being pushed forward beyond the Old Stage road, and, on the right flank, vedettes occupied a hillock near the mouth of the creek. One Whitworth gun was posted on the heights beyond the Massaponax, north-east of Yerby's House.

The entire length of Lee's line was about 11,500 yards, and his combatant strength being probably about 68,000,[34] the proportion of men to space was 11,000 to the mile, nearly six men to the yard. His left was naturally strong, and the fords above Falmouth were watched by cavalry; this portion of the position, therefore, was held by one division only. The salient of Marye's Hill, commanding the two roads, was manned by two divisions, and its natural defensive capabilities artificially improved. The re-entrant of the centre, flanked as it was by bastion-like spurs, was thinly occupied by 13,000 men. On the right, the weaker flank, where the greater width of the plain gave the enemy room to deploy, and where defeat would cut the army off from the railroad, and perhaps from the Telegraph road, was massed a whole army corps, with a strong cavalry division operating on the flank.

34. Two cavalry brigades, Rosser's and Hampton's, about 5,000 men, were detached at a distance from the field of battle.

Lee is reported to have expressed his regret after the battle of the 13th, that his line had not been more thoroughly prepared for defence; and it might be supposed that this oversight caused him some anxiety during the course of the engagement. The contrary, however, was the fact; probably no general or army ever awaited the attack of a more numerous enemy with greater confidence than did Lee and the Confederates at Fredericksburg. Almost every precaution had been taken that skill suggested and time allowed; there was but one weak link in the chain of defence, the projecting coppice, and it may be that this was in Lee's mind when he spoke of incomplete preparation. Shortly after eight o'clock on the morning of the 13th, accompanied by his slender personal staff, he rode along the front; and having satisfied himself that all was in order, returned to the centre, and took his stand with Longstreet upon that wooded height which has since borne the name of Lee's Hill.

We will now consider the dispositions for attack made by the Federal commander, and in order to understand his action, it must be held in mind that he still believed, although Lee had received forty-eight hours' warning of his design, that the Confederate right wing was in position near Port Royal, and that he was confronted by only a portion of the enemy's force.[35] As to his actual plan of battle there is a conflict of evidence, which is, to say the least, curious, but at the same time scarcely worth discussion. As far as can be elicited from the mass of contradictory testimony recorded, his final decision, delivered at an informal council of war, held on the night of the 12th, was to make his principal effort against the enemy's right, his weakest point, and to support it by an assault of Marye's Hill on the left centre. So far, so good; the deep channel and marshy valley of the Massaponax made a turning movement exceedingly hazardous, and the plan under the circumstances was sound enough. Marye's Hill was certainly too strong to be carried by a direct attack, but it was understood that operations against it would be confined to a brisk demonstration until an opportunity for assault should present itself.

35. This was actually the case at sundown on the 12th; but the Federal Intelligence Department seems to have failed to ascertain, or at any rate to report, if it was so on the following morning.

If, after communicating his scheme to his subordinates, Burnside had been content to trust to them the execution thereof, it would have been better for his reputation. The commanders of the three Grand Divisions, on the night of the 12th, after the council of war, were under the impression that the above plan was to be carried out to the letter, and that final orders as to the hour of attack would reach them early the next morning. During the night, however, Burnside, inexperienced in command, wavered from his purpose, and committed himself to half-measures. By the council of war the attack of the enemy's right was committed to Franklin, and in addition to his own Grand Division, two divisions of the Third Corps, of Hooker's command, were placed at his disposal.

Sumner was in charge of the operations against Marye's Hill, reinforced by Whipple's division of the Third Corps.

Franklin had under his orders eight divisions and 116 guns: 55,000 men.

Sumner had under his orders seven divisions and 60 guns: 30,000 men.

The Fifth Corps remained under Hooker on the far side of the river as a general reserve of three divisions and 30 guns: 19,000 men.

The ordnance on the Stafford Heights numbered about 120 guns; sixty-one, under De Russey, were posted opposite the Confederate right, the remainder above the town and along the heights beyond Falmouth.

Impatiently on the morning of the 13th Franklin awaited his orders, but when after long delay they at last arrived, he found to his surprise and chagrin, that, instead of receiving carte-blanche, as he expected, to use his whole command against the Confederate right, he was fettered by instructions to employ an attacking column of "not less than a division," and at the same time to hold his troops in readiness to move rapidly down the Old Stage road. Nor had the two promised divisions of the Third Corps been as yet instructed to join him.

What was he to infer? Was the attack and defeat of the Confederates to be left to Sumner alone, while he himself, covering his movement by seizing the heights with a portion of his force, prepared to turn against the enemy's wing erroneously supposed to be still at Port Royal?

If a serious and determined attack was intended, why was any limit placed on the numbers to be employed? However weak the Confederates might be upon their right, it was certain that to wrest from them the line of heights, and to inflict upon them a crushing defeat, was scarcely within the power of a single division of 6,000 or 7,000 bayonets. On the other hand, there was no suggestion that the original scheme had been departed from, and that the main attack would be undertaken by Sumner. The wording of the despatch was certainly vague and unsatisfactory, and the meaning was difficult to fathom. Burnside, however, did not leave his subordinate without an interpreter. Before the battle commenced his Chief of the Staff joined Franklin and remained with him throughout the day. But unfortunately the indecision and vacillation of the general-in-chief had become too apparent. Harassed by doubts, and without confidence in his superior, Franklin in his turn resorted to half-measures, than which, in warfare, nothing is more hopeless. It cannot be too often repeated that want of determination in the execution is fatal to any plan of action, however well-conceived: a weak plan, boldly carried out, is far more likely to be successful than one sound in itself but prosecuted without vigour and resolution.

Sumner on the right also received orders which limited the numbers of the assaulting force to one division only. Knowing what we do of the strength of the point to be attacked, and bearing in mind that it was apparent to Burnside also, we must convict him of a capital error in directing the assault to be undertaken with an inadequate force; or, if he meant the assault only to follow demonstration, and to be dependent on the success of the left wing, in not clearly expressing his intention.

With the vague and misleading instructions of the inexperienced Federal general, readers of military history will compare the clear and precise orders of battle issued by Napoleon, Wellington, or the German generals of 1870–71; the minute directions and explanations communicated even to the rank and file by the great Russian leader, Skobeleff; or the practice of Burnside's opponent, Lee. The Southern commander, having brought his troops up to the position he had selected, communicated without reserve to his lieutenants his general plan, and then gave frankly into their hands the conduct of the fight.

The lesson should be impressed upon all officers who may have the direction of any military operation, however insignificant. Make up your own mind clearly as to the course you intend to pursue. Let your plan be as simple as possible; well-defined both as to the means to be employed and the objective aimed at: take care that it is prosecuted with the utmost determination, and, if your subordinates are qualified to command, leave them to themselves, and beware of unnecessary interference. Be determined at the same time, if necessary, to enforce prompt co-operation towards the end in view. At Gettysburg, Lee's first defeat, a want of determination on the part of the great Virginian general in compelling exact and timely obedience to his orders, is believed by many to have lost the battle.

Let your instructions be explicit, plainly-worded, and capable of no double construction; and, above all, see that in letter and in spirit they are understood by those upon whom you depend to carry them out. Time employed in explaining orders is never thrown away: let all ranks understand what is required of them, and you will ensure the intelligent co-operation, not only of the officers, but of your whole force.

The position of the Federal army on the night of the 12th and the following morning will be best comprehended by reference to the map. [Map IV.]

Shortly after nine o'clock on the 13th, the sun, shining out with almost September warmth—for though the year had nearly reached its close, the pleasant Indian summer still lingered among the woodlands of Virginia—quickly dispelled the mists which hid the opposing armies; and as the dense white folds dissolved, Jackson's men beheld the plain beneath them dark with a moving mass of more than 40,000 foes. Franklin's Grand Division was in motion on the Old Stage road: down the face of the cliffs beyond the river, and across the narrow bridges, poured an unceasing stream of men and guns; and from the long array of batteries upon the Stafford Heights a great storm of shot and shell burst upon the Confederate lines.

And yet that vast array, formidable in numbers, training, and equipment, lacked the moral force without which physical power, even in this its most terrible form, is but an idle show. The dissensions of the leaders, the want of energy and decision in carrying out the preliminary movements, the insecurity of their situation, were but too apparent to the intelligence of the regimental officers and men. Northern writers have recorded that the Army of the Potomac never went down to battle with less confidence and alacrity than on this day of Fredericksburg.

Hidden by thick timber, crouching behind their parapets, or ensconced in deep ravines, Lee's soldiers lay secure, undisturbed by the tempest that crashed harmlessly above them through the leafless branches; and, reserving their fire for the hostile infantry, the guns were silent.

How dark and forbidding must that long line of hill and wood, manned by unseen foes and wrapped in silence more terrible than the fiercest din of battle, have appeared to the advancing Federals! To their credit be it said, the Northern infantry never attacked more resolutely than they did at Fredericksburg; and, had the capacity of the leaders been equal to the courage of the men, Lee's veterans would have been hard set to hold their own.

Left Attack.

[9.30 a.m.] About half-past nine, Franklin, who had moved across Deep Creek and down the Old Stage road for nearly half a mile, fronted, and advanced against the heights.

In accordance with Burnside's order the attacking line consisted of one division only, the Third of the First Corps, under Meade, composed of Pennsylvanian regiments. The Second Division of the same corps was in support on the right under Gibbon, and the First, commanded by Doubleday, was placed *en potence* on the left along the Mine road.[36] Franklin was evidently apprehensive of a counter-attack against his left; otherwise the First Division would hardly have been retained in this position, as it was throughout the day.

The Sixth Corps[37] held the Old Stage road on either side of Deep Run; two divisions in first line and a third in support. A grand battery of thirty-six pieces, stationed 600 yards from the

---

36. Approximate strength:—Meade, 4,600; Gibbon, 5,500; Doubleday, 7,000. ["Troops are ranged *en potence* by breaking a straight line, and throwing a certain proportion of it . . . backward . . . for the purpose of securing that line. . . . The disposition *en potence* is frequently necessary in narrow and intersected ground" (Charles James, *A New and Enlarged Military Dictionary* . . . [London, 1805]).]

37. 21,000 strong.

river-bank, served as a *tête-du-pont*. The numerous field guns were placed along the front in the intervals between brigades, the batteries acting independently.

Meade's first brigade, covered by one battalion skirmishing, was followed at a distance of 300 paces by the second, also in line; the third, marching in column of fours, moved a little to the left rear of the first, with one regiment thrown out in extended order on the flank. The presence of Stuart's dismounted troopers away to the left evidently dictated this formation.

The skirmishers in front were soon briskly engaged with Hill's sharpshooters, but no sooner had the compact line of the leading brigade crossed the Old Stage road and gained the crest of a shallow hollow which lay between it and the woods, than the left was assailed by a well-directed and raking artillery fire.

Captain Pelham, who commanded the Confederate horse artillery, galloping forward with two guns as the Federals advanced, and escorted by a dismounted squadron, had come into action near the cross roads, enfilading the enemy's line. So telling was his fire that the first brigade wavered and gave ground; and though Meade quickly brought up his guns and placed his third brigade *en potence* in support of them, he was unable to continue the advance until he had brushed away his audacious antagonist. The four divisional batteries were aided by two of Doubleday's; but, rapidly changing his position as often as the Federal gunners found the range, for more than half an hour the daring Southerner defied their efforts, and for that space of time arrested the advance of Meade's 4,500 infantry. One of his pieces was soon disabled, but with the remaining gun, a twelve-pounder Napoleon, taken from the Federals six months before, he maintained the unequal fight until his ammunition was exhausted, and he received peremptory orders from Stuart to withdraw. That the skirmish line of Meade's third brigade was not used to drive away this intrusive assailant, as would now be the case, is probably due to the fact that at the time skirmishers never attacked on their own account, but were used merely to cover the movements of the main body.

This gallant action made the name of the young soldier—he was but two-and-twenty years of age—one of the most famous in the Confederacy.

[11 a.m.] On Pelham's retirement, Franklin, bringing several batteries forward to the Richmond road, for more than half an hour subjected the woods before him and the line held by the dismounted cavalry to a heavy cannonade; and at the same time demanded from Hooker the two divisions of the Third Corps[38] which Burnside had placed at his disposal. It was past eleven before Meade recommenced his advance. Covered by the fire of the heavy guns upon the Stafford Heights and of field batteries on either flank, his line had gained a point within 800 yards of the foot of the opposing ridge, when suddenly the silent woods awoke to life, and the flash and thunder of more than sixty guns revealed to the Federals the magnitude of the task they had undertaken.

Walker's fourteen guns on Prospect Hill, Stuart's eighteen beyond the cross-roads, and the great batteries on the left of Hill's division, opened almost simultaneously.

The skirmishers were quickly driven in, and on the closed ranks behind burst the full fury of the storm. Dismayed and decimated by this fierce and unexpected onslaught, Meade's brigades broke in disorder and fell back to the shelter of the road.

For the next hour and a half an artillery duel raged in this quarter of the field: the Confederate guns, their position at last revealed, engaged with spirit the more numerous and powerful batteries of the Federals; while Franklin brought up three fresh divisions to support his foremost line.

About twelve o'clock, part of D. Hill's division advanced in support of the Southern cavalry, but was almost immediately withdrawn.

Right Attack.

The original design, approved by the Federal council of war, had been to make the main effort on the left, and to confine the right wing to demonstration until a favourable opportunity to convert a feigned attack into a real one should present itself. We shall now see how Burnside, having already, by useless limitations as to the force to be employed, fettered and disheartened his lieutenants, was himself the first to depart from

38. Birney's and Sickles's, about 14,000 men.

the proposed line of action, and to turn awry the current of the enterprise.

At eleven o'clock, as Meade on the left was preparing to advance for the second time, Sumner received instructions to move out to the assault of the heights before him. At the same time Longstreet's guns opened fire, and enfiladed the bridges and streets of the town to such purpose, that the Federal columns had to creep forward in single file, clinging to the walls and buildings on either hand.

The Second Corps occupied Fredericksburg; the Ninth lay south of the town on the far side of Hazel Run; Whipple's division of the Third Corps was in reserve. Sumner had under his orders sixty guns, but not more than eight of his batteries were able to come into action.

It was about twelve o'clock when French's division of the Second Corps, 4,500 strong, which had been deputed to carry out the assault, issued from the town by the plank and Telegraph roads.

At this moment, Meade, as we know, had not only failed to make the slightest impression on the enemy's right, but had actually suffered a decided check. His movement had, however, served the purpose of a reconnaissance in force, and it must have been clear, even to Burnside, that Jackson's corps was present, and the whole Confederate army concentrated on the heights before him. The Federal general, however, notwithstanding the repulse of the left attack, did not recall his instructions issued to Sumner at eleven o'clock, but allowed that commander to make an isolated attempt against the most formidable point of Lee's line, unsupported by any combined movement of the left wing, which at the time was bombarding the woods, but making no effort whatever to engage the enemy's infantry.

There was little preparation for the attack on the part of the artillery. The heavy guns on the Stafford Heights had been but for a short time in action. The dense mist, which lifted about nine o'clock from the lower reaches of the stream, lingered long over the ridge which stands above the town; and the Confederate batteries on the open plateau, visible only at intervals, had as yet escaped disaster.

[12 noon.] French's division, covered by the usual screen of skirmishers in front and on the flank, and supported by six batteries posted on the outskirts of the town, advanced along the roads in column of fours, the enemy's picquets retiring slowly before it; and, though exposed to the fire of Walton's and Maurin's thirteen guns and of the batteries on Stansbury and Taylor's Hills, it crossed the mill-sluice by the bridges, and mounted the slope of the miniature valley through which it runs. In the ploughed fields beyond, the leading brigade attempted to deploy, but ere the movement was two-thirds completed, scourged by musketry and by volleys of round shot, which burst wide gaps in the ranks that were visible a mile away, the column fell back to the town, having lost 1,200 men. Three camp colours, left standing 175 yards distant from the foot of Marye's Hill, marked the extent of the front of the first brigade and the limit of the advance.

[12.30 p.m.] Fifteen minutes passed and another division, Hancock's, 5,000 strong, rushed forward from the town. Zook's brigade led the way, but quickly recoiled, beaten back by that terrible artillery. Not so its successor. Under cover of the further bank of the ravine, the Irish Brigade,[39] under General Meagher, threw off their haversacks and blankets and deployed into line. Resolutely the 1,200—for they were no more—breasted the slope and faced the death-dealing storm: swiftly they passed the limit marked by the three solitary colours, and shoulder to shoulder, their own green flag and the blue and scarlet of the Union standard waving above them, swept forward against the low wall which skirts the base of Marye's Hill.

So determined was their advance, that Colonel Miller,[40] commanding the Confederate brigade confronting them—for General Cobb had already fallen—ordered his men to hold their fire for a space. And now occurred a strange and pathetic incident. Though high was the courage of that thin line which charged so boldly across the shot-swept plain, opposed to it

---

39. Composed of the 28th and 29th Massachusetts, the 63rd, 69th, and 88th New York, and the 116th Pennsylvania. [The 29th Massachusetts was not a part of the Irish Brigade but belonged instead to the Second Brigade, First Division, Ninth Army Corps (*Official Records*, XXI, 52).]

40. [The correct name of the officer who succeeded temporarily to the command of Cobb's brigade was Colonel Robert McMillan (*ibid.*, p. 607).]

The Campaign of Fredericksburg

were men as fearless and as staunch: behind that rude stone breastwork were those who were "bone of their bone and flesh of their flesh;" the soldiers of Cobb's brigade were Irish like themselves. On the morning of the battle General Meagher had bade his men deck their caps with sprigs of evergreen, "to remind them," he said, "of the land of their birth:" the symbol was recognized by their countrymen, and "Oh, God, what a pity! Here come Meagher's fellows," was the cry in the Confederate ranks.

One hundred and fifty paces from the hill, the brigade halted and fired a volley, while the round shot tore fiercely through the ordered line. Still no sign from the wall, looming grim and silent through the battle-smoke; and again the battalions moved swiftly forward. They were but a hundred yards from their goal, unbroken and unfaltering still; they had reached a point where Walton's gunners, unable to depress their pieces further, could no longer harass them. Victory seemed within their grasp, and a shout went up from the shattered ranks. Suddenly, a sheet of flame leaped from the parapet, and 1,200 rifles, plied by cool and unshaken men, concentrated a murderous fire upon the advancing line. To their glory be it told, though scores were swept away, falling in their tracks like corn before the sickle, the ever-thinning ranks dashed on,

> "The charging blood in their upturned faces,
> And the living filling the dead men's places."

But before that threatening onset the Confederate veterans never quailed; volley on volley sped with deadly precision, and at so short a range every bullet found its mark. For a while the stormers struggled on, desperate and defiant; but no mortal men could long face that terrible fire, scathing and irresistible as the lightning, and at length the broken files gave ground. Slowly and sullenly they fell back; fell back to fight no more that day, for beneath the smoke-cloud that rolled about Marye's Hill the Irish Brigade had ceased to exist. Of 1,200 officers and men, 937 had fallen.[41] Forty yards from the wall, where the

---

41. [Henderson's casualty figures are exaggerated. Meagher reported that "out of 1,200 men I had led into action that morning about 250 alone had reported to me under arms from the field" (*ibid.*, p. 243). But according to the official return of casualties that was compiled later, Meagher's brigade lost only 545 officers and men. *Ibid.*, p. 129. This is the generally accepted figure.]

charge was stayed, the dead and dying lay piled in heaps, and one body, supposed to be that of an officer, was found within fifteen yards of the parapet.

The adjutant-general of Hancock's division, who witnessed the attack from the town, said that at the time he could not understand what had happened; the men fell in such regular lines that he thought they were lying down to allow the storm of shot to pass over them. General Ransom, commanding one of the divisions which held Marye's Hill, reported that this assault was made "with the utmost determination;" and the eloquent words of the *Times* Special Correspondent, who was present with the Confederates, record the admiration of those who beheld that splendid charge: "Never at Fontenoy, Albuera, or Waterloo, was more undaunted courage displayed by the sons of Erin; the bodies which lie in dense masses within fifty yards of the muzzles of Colonel Walton's guns are the best evidence what manner of men they were who pressed on to death with the dauntlessness of a race which has gained glory on a thousand battle-fields, and never more richly deserved it than at the foot of Marye's Hill, on the 13th day of December, 1862."

After the battle, on the ground over which the divisions of the Second and Third Corps had passed, and within a space not larger than two acres in extent, 680 corpses were counted, lying in many places literally in heaps; and it was noticed that the faces of most of them were of the Milesian type. This spot was significantly named the Slaughter-pen. Two hundred paces in rear of Meagher's line, Caldwell's brigade had moved forward, but, disheartened doubtless by the fate of its gallant predecessor, was more easily repulsed; and when the relics of the Irish regiments had been driven from the frail shelter of the fences and wooden buildings where they had taken refuge, the Confederate fire ceased. Hancock's division, which had gone into action 5,006 strong, lost 156 officers and 2,013 men. Six Confederate regiments and about thirty guns were actively engaged. Twenty minutes only elapsed from the moment Zook attempted to deploy until the broken and bleeding remnants of Meagher's and Caldwell's brigades reeled back to the bank of the mill-sluice.

Colonel Miller's [McMillan's] bold action in ordering his

infantry to suspend their fire, at the very moment the advance appeared most formidable, cannot be too highly commended. Far more destructive to the *morale* of an attacking force is the sudden shock of well-directed volleys delivered at close quarters, than continuous exposure to fire commenced at long range and maintained as the interval decreases, the loss from which is usually slight in proportion to its intensity. Nor does a protracted fire-action tend to steady the nerves of the defenders: men who have been using their rifles for some time are apt to get excited and careless; and at the same time to lose confidence in their power of resistance, at the sight of the enemy advancing despite their efforts.

To keep your men under cover as long as possible and to open fire only at the most effective range, the point-blank, is, when acting on the defensive and supported by artillery, sound tactics. But to do this effectually troops must be not only securely covered but also thoroughly trained and well in hand; and it speaks volumes for the fire-discipline and steadiness of the Confederate infantry that they were able to carry out the order with such coolness and effect. General Kershaw, who took over the command along the wall soon after Cobb's death, reported that though at one time his line was in parts four deep, and the men, kneeling to load, rose by ranks to deliver volleys, not a single soldier was injured by a comrade's fire.

On the Confederate side, as French's division debouched from the town, Cooke's brigade of 1,600 men was brought up from the reserve slope of Marye's Hill and placed in line before the house and garden; and, as the Irish Brigade deployed, the two regiments on the right advanced to the crest and aided in repulsing the attack.

It was now nearly one o'clock. Two divisions had been hurled fruitlessly against the hill, 10,000 Federals had been easily and bloodily repulsed by a force one-fourth their number, and it seems almost incredible that Burnside should have allowed, without changing his tactics, another division to be swallowed up in that gulf of fire. The battle was to be won, if at all, by crushing the enemy's right. His centre and left were not only naturally strong, but were adequately defended: moreover, success there would be but partial, for the open plateau was commanded by the further ridge. Franklin's first attempt had failed, it is true, but it had been made with inadequate force, and there was no reason why a stronger effort should not succeed. There was no doubt that on the right, the tactical flank, the Confederates were in strength, and to defeat them it would be necessary to employ every available man and gun. What Burnside ought to have done was to have reinforced Franklin as strongly as possible, and, refusing his right, to have confined Sumner as strictly to demonstration as was compatible with the security of that wing and of the bridges. As usual, he did exactly the contrary. He sent the whole Fifth Corps of Hooker's Grand Division to assist Sumner, and to Franklin but a single division, the first of the Ninth Corps.

[1.15 p.m.] About 1.15 p.m., Howard's division, in the same formation and on the same narrow front as French and Hancock, supported by one solitary brigade (of Whipple's division) on his right rear, advanced from the lower quarter of the town; and again the wave of attack broke vainly against that fatal hill.

Kershaw, as the Federals came forward, reinforced the centre of the line with two of his own regiments, held hitherto in reserve in the ravine of the Hazel Run. These were doubled on the Phillip's Legion and the 24th Georgia. One of Ransom's regiments was brought up on the left of the Telegraph road lining the crest in rear of the 24th North Carolina, while the remainder of the brigade took post in close support.

Howard lost 877 officers and men.

Left Attack.

[1 p.m.] At one o'clock, Franklin, having silenced the Confederate guns, again essayed the task of carrying the heights. Gibbon's division had been already posted on Meade's right. Birney's (Third Corps), which crossing the bridge at ten o'clock, had arrived on the ground at noon, was in support of the left; Newton's some distance to the right rear.[42] Brook's and Howe's maintained their former position on either side of Deep Run. Sickles's division (Third Corps), which had followed Birney's, was retained in reserve near the bridges; and Doubleday's, still *en potence* on the left, held Stuart's cavalry in check. Twenty-

42. Each division about 7,500 strong.

The Campaign of Fredericksburg 77

one guns on the right and thirty on the left of the attacking line, stationed on the road, 1,000 yards from the enemy's position, supported the movement. Preceded by clouds of skirmishers and covered by a tremendous artillery fire, Meade and Gibbon advanced from the Old Stage road in the usual formation, column of brigade at 300 paces' distance, the whole covering a front of 1,000 yards; whilst Birney replaced Meade along the road.

When the Federals reached the scene of their former repulse, Jackson's guns again opened; but without the same effect, for they were now exposed to the fire of the enemy's more powerful artillery. Even Pelham could do but little, and the batteries which had been advanced beyond the railroad on Hill's left front were quickly driven in;[43] the supporting regiment advancing to a little hillock twenty yards beyond the embankment and firing volleys to check the enemy's skirmishers and cover the retirement.

Meade's rearmost brigade was now brought up and deployed to the left of the first, thus further extending the front.

The leading brigade made straight for that tongue of woodland, which, projecting beyond the railroad, interposed between Lane and Archer. As they approached the battalions found that, masked by the timber, they were no longer exposed to the fire of the enemy's artillery, and that the wood before them was unoccupied.

Quickly crossing the border, through swamp and undergrowth deemed impenetrable by the Confederates, an ever-increasing stream of men pressed on, and bursting from the covert to the right hand, attempted to roll up the exposed flank of Lane's brigade. The ground between the northern skirt of the wood, however, and Lane's right was an open field, 200 yards in width, and over this space the Federals made no immediate progress. At length, their ammunition giving out, the Southerners retired, but slowly and in good order. Neither Gregg nor Archer were able to lend assistance, for they themselves were fully occupied with Meade's second brigade; which, though following close on the heels of the first and met at the entrance to the coppice by the oblique fire of Lane's regiments,

43. These guns joined the great battery posted near Bernard's Cabin.

had, instead of conforming to the change of direction, rushed forward through the wood.

Two hundred paces from the embankment it came in contact with Archer's left, which was resting on the very edge of the coppice. The Confederates were completely taken by surprise: relying on the reported inaccessibility of the thickets beyond their flank, neither scouts nor picquets had been thrown out to watch the approaches, and the men were lying about with arms piled. Two regiments, attempting to form up and change front to the left, were assailed in the act, broken by a determined charge, and routed in confusion. The remainder, however, stood firm, without changing front, for the Federals, instead of following up their success in this direction, left Archer to be dealt with by the third brigade of the division (which had now reached the railroad and was threatening him in front), and swept on towards the military road, where Gregg's brigade was drawn up in line. It would have been well had a portion at least remained to assist, by a flank attack, in forcing Archer back, for the third brigade, left to itself, proved unequal to the task.

So thick was the covert, and so limited the view, that General Gregg, taking the advancing mass for part of Hill's line retiring, restrained the fire of his men. The Federal rush broke upon his right. He himself fell mortally wounded; his flank regiment, a battalion of conscripts,[44] fled, except one company, without fir-

44. [Like most British soldiers of his day, Henderson was opposed to the idea of conscription. He interpreted the Civil War, and later the war in South Africa, as "a triumph for the principle of voluntary service" (*The Science of War*, p. 379). Yet the facts in this case do not reveal—as Henderson seemingly implies—that Gregg's right regiment broke because it contained a large element of conscripts. Chesney, too, mentioned "one regiment of raw Carolina conscripts" that had abandoned its intrenchments (Charles Cornwallis Chesney, *Campaigns in Virginia, Maryland, ...* [London, 1864–65], I, 187), but Freeman refers to the organization in question, Orr's Regiment of Rifles, as "veterans" (Douglas Southall Freeman, *Lee's Lieutenants* [New York, 1944], II, 355). Actually this regiment was ordered by General Gregg to "refrain from firing" to prevent injury to Hill's men retreating from the first line, and before the men knew it "the Federal line was right upon the Rifles. . . . Then ensued a scramble and hand-to-hand fight. . . . The rifle regiment was, as a body, broken, slaughtered and swept from the field" (J. F. J. Caldwell, *The History of a Brigade of South Carolinians, Known First as "Gregg's," and Subsequently as "McGowan's Brigade"* [Philadelphia, 1866], p. 59). In his report, A. P. Hill remarked that the conscripts in the Light Division "showed themselves desirous of being thought worthy comrades of our veteran soldiers," and one of

ing a shot. The two regiments on the opposite flank, however, were with great readiness turned about, and changing front inwards, effectually obstructed any movement of the enemy along the rear. But the Federals, though now joined by part of the first brigade, had already reached the limit of their success. The Pennsylvanian regiments found themselves in the heart of the enemy's position; but from the very nature of their advance and of the ground over which they had passed, they had become a confused and disorganized mass, and to them may be well applied the words of Kinglake:—"They were only a crowd, and they all of them, simple and wise, now began to learn in the great school of action that the most brilliant achievements of a disordered mass of soldiers require the steady support of formed troops; then, as it is said, for the first time the men cast a look toward the quarter from which they might hope to see supports advancing."

Let us now look at the progress and position of the different bodies from which the much-needed aid was expected.

To the right rear, opposite Pender, Gibbon's division, which had moved to the attack almost simultaneously with Meade, had been checked by the fire of the thirty ppo Confederate guns posted *en masse* 300 yards behind the railroad. Two of his brigades had been driven back; the third had with difficulty gained the shelter of the embankment.

To the left rear, Meade's third brigade was held in check by Walker's batteries and the staunch infantry of Archer, who, notwithstanding that a strong force had passed beyond his flank, resolutely held his ground, and prevented his immediate opponents from reinforcing the intruding column.

Not a single field battery had followed in support of the infantry; between the railway, therefore, and the Old Stage road, a distance of 950 yards, there was no body of formed troops. Meade, with less than 3,000 men, was alone and unsupported within the hostile lines, swallowed up by the forest and surrounded by an overwhelming throng of foes. At this crisis of the fight, when every available battalion should have been hur-

---

Hill's brigadiers reported that the conscripts, many of whom were good soldier material, had "proved themselves worthy" (*Official Records*, XXI, 647, 656; Bell Irvin Wiley, *The Life of Johnny Reb* [Indianapolis, 1943], p. 342).]

ried to the front and poured through the still open gap, when a determined rush of the whole fighting line and supports would have probably driven Hill and Early back upon the reserves, Franklin, incapable of a bold offensive, made no effort to assist his lieutenant, and, despite three urgent appeals for succour, left the gallant Pennsylvanians to their fate.

If Birney, generously responding to Meade's cry for help, had, with Newton on his right, swept swiftly across the open, and, overwhelming Lane and Archer, had pressed on to the military road; if Doubleday, abandoning the passive defensive, had threatened Stuart and induced Jackson to detach to the aid of the cavalry a portion of his reserve, all might yet have been well. It was not to be, however; timidity or incapacity neglected the opportunity. Franklin, holding in his hand 40,000 infantry at least, saw those daring troops who, though numbering scarce 3,000, had so successfully cleared the way, destroyed piecemeal by his own violation of the first principles of war.

Whether, if he had supported Meade and Gibbon with 20,000 bayonets, he would have hurled Jackson back, is another matter; still he himself would have been quit of blame. Lee on other fields had shown himself a master of the delicate operation of thinning one part of the line to strengthen another; in addition to D. H. Hill and Taliaferro, Hood's and Pickett's reserve brigades were not far distant, and the soldiers of the Antietam could boast that the iron front of their resistance had never yet been broken through.

The Confederates, in sharp contrast, and prompted, perhaps by the urgency of the case, were not slow to come to the assistance of their comrades. Early, anticipating Jackson's orders, hurled the brigades of Atkinson and Walker against the flank of the hostile column and despatched Hoke, with Hays in support, to reinforce Archer. At the same time Thomas, supported by Paxton, charged forward to the relief of Lane; and the combined brigades bore back the Federals, outnumbered but fighting stubbornly, down the slope, and after a brief yet desperate struggle, thrust them from the woods.

Archer also, promptly appreciating the situation, detached four regiments from his right, led them along the rear of the battalion that still faced the front, and dashed upon the struggling mass. Though compelled by sheer force of numbers to retreat, Meade's men still showed a bold front, and on gaining the railway embankment turned fiercely to bay. But in the thick covert they had been thrust from, all order and cohesion had been lost, and ere they could make good their grip upon that line of vantage, the Confederates rushed down upon them with the bayonet, and drove them far across the plain.

[2.15 p.m.] Walker, Thomas, and Archer, halted at the embankment, for four regiments, sent too tardily by Birney to Meade's aid, were attempting to enter the projecting coppice; but Atkinson and Hoke, carried away by success, pursued the fugitives beyond the railroad.

In Atkinson's path stood a field battery, abandoned by the gunners as the Confederates approached. The Georgian regiments, their ammunition already exhausted, were within eighty yards of the cannon, when the first line of Birney's division, drawn up on the Old Stage road and supported by sixteen guns, shattered their right with sudden volleys of musketry and canister, arrested their progress, and snatched away the anticipated prize. The brigade fell back, leaving the brigadier, his adjutant, and over 300 men upon the field. Before Hoke's onset, Birney's left gave way for fifty yards or more; but the Confederates, in their turn unsupported, were unable to follow up their success; and after suffering heavy loss, withdrew from the exposed situation where their reckless impetuosity had placed them.

In killed, wounded, and prisoners, Meade lost more than 2,000 officers and men. When Meade gave way, Gibbon's troops, who, it has been suggested, were of inferior quality, and had not, in the face of the powerful artillery that confronted them, been able to gain ground beyond the railroad, scattered over the plain in headlong flight.

It should be mentioned that when the Confederate batteries, thrown forward beyond the railroad, were driven in by Gibbon's advance, one division [battery?], instead of retiring, took post several yards further to the left front, whence it enfiladed the Federal line with great effect, and probably did much towards preventing Gibbon's brigades firmly establishing themselves upon the enbankment.

Birney, though he responded too late to Meade's demand for

succour, did good service during the repulse. His four regiments sent forward to assist Meade's retiring soldiers in their last attempt to rally, became involved in the rout, and lost 600 officers and men. As the fugitives of the First Corps came flying towards the road, he deployed two regiments across their path, but was unable even by such means to stay the career of the defeated divisions.

We have already seen how he beat back with heavy loss the isolated counter-stroke of Hoke and Atkinson. In his report he claims, by thus checking the pursuit, to have saved the left wing of the army from overthrow. However this may be, his words, speculating on the probable disastrous effect of a general counter-attack at this moment, bear strong testimony to the extent of Jackson's success, and of the Federal demoralization.

Meade and Gibbon lost together 4,000 men, exactly two-fifths of their number, but secured at the same time 200 prisoners.

To recapitulate:—10,000 infantry had attacked a strong position held by unknown numbers; two brigades had penetrated the enemy's line, but when, unsupported, they had been compelled to retire, the whole attack had given way. The artillery of the defence had been silenced—not crushed—by the preparatory fire of the enemy's batteries, and re-opening at the opportune moment, had assailed, with terrible effect, his advancing infantry. Three Confederate brigades, Archer, Lane, and Atkinson, had suffered somewhat severely, losing 1,600 out of 5,800.[45]

Franklin, when his first line broke, had still intact the divisions of Birney, Doubleday, Sickles, and the entire Sixth Corps, 42,000 men in all; and at least one division of the Ninth, and one of the Fifth Corps (which had crossed the river about two o'clock) might have been sent to his assistance in case of need. Thirty thousand infantry, of which not more than one-third had been engaged, were, together with the cavalry (4,000), available on the Confederate side for offensive movement. Two of Franklin's divisions, however, Howe and Brooks, who had

45. [This statement is misleading in that the figures given represent the casualties of A. P. Hill's entire division. The brigades of Archer, Lane, and Atkinson suffered slightly better than 1,000 casualties (*Official Records*, XXI, 560–61).]

suffered heavily from artillery fire, were occupied in containing Hood and Pickett, and four only (for Meade's and Gibbon's troops had been badly cut up) were at hand to resist an immediate attack: that is, if Jackson had fallen on the Federals, he would have been met by but 25,000 bayonets, of which 7,000 (Doubleday's division) were posted at right angles to his line of attack. Supporting troops would have doubtless been sent over from the Federal right, but, be it remembered, Franklin's masses were bound to a narrow and hazardous line of retreat, and had already been witnesses of the demoralizing repulse of the attacking column.

If the Confederate general-in-chief had ordered Jackson and Stuart, together with Hood and Pickett, to close with the enemy, there appears no valid reason why the whole left wing of the Federal army should not have been driven with fearful slaughter into or beyond the Rappahannock.[46]

However, the Confederates, ignorant as they necessarily were of the mistrust and want of confidence in its leaders with which the Federal army was infected, were unaware of the extent of the demoralization of which this feeling was a cause, and were far from suspecting what a strong ally they had in the very hearts of the enemy. Moreover, so easily had the attacks been repulsed, that both Lee and Jackson never doubted but that they were merely tentative efforts, the prelude, as it were, to the main assault.

To return to Franklin. As soon as the pursuit ceased, Birney and Sickles were ordered to maintain the position held before the attack by Meade and Gibbon. Newton still kept his place in the centre; and Burns's division of the Ninth Corps,[47] which, despatched from Sumner's right, reported to Franklin at three

---

46. [Additional research later prompted Henderson to modify this opinion. In 1898 he wrote: "General Lee's arrangements . . . had not included preparation for a great counterstroke, and such a movement is not easily improvised. The position had been occupied for defensive purposes alone. There was no general reserve, no large and intact force which could have moved to the attack immediately the opportunity offered" (*Stonewall Jackson and the American Civil War* [new impression; New York, 1906], II, 321). Freeman agrees that the chances for a successful counterattack, even on Jackson's front, were not promising (*Lee's Lieutenants*, II, 369). In his biography of Jackson, Henderson leaned heavily upon *The Campaign of Fredericksburg* in describing the battle; often he quoted substantially from the earlier work.]

47. Probably not more than 3,500 bayonets.

o'clock, supplied the place of Sickles in reserve. The broken divisions of Meade and Gibbon were reformed along the riverbank, south of Deep Run.

[2.30 p.m.] While these movements were in progress, at 2.30 p.m. Burnside, having just witnessed the signal failure of a fourth attempt to carry Marye's Hill, sent an urgent order to Franklin to renew the attack. Beyond informing the general-in-chief that fresh troops (Sickles's division) had been sent in, Franklin made no response; he had lost all confidence in his superior, and took upon himself to disobey; and indeed, had he ordered an advance, his soldiers, disheartened and exasperated by the feeble generalship which had so recklessly and unavailingly sacrificed their comrades, would in all probability, as they did under Grant at Cold Harbour in 1864, have refused to move.

On the Confederate side Taliaferro and Early, with part of the Light Division, now held the railway embankment and the skirt of the woods; D. H. Hill was brought up in support, and the shattered brigades of A. P. Hill were withdrawn behind the Mine road. During the rest of the afternoon the skirmishers were actively engaged, but though Jackson's victorious soldiery long and eagerly awaited a renewal of the assault, the disheartened Federals refused to be again tempted to close conflict. Jackson had lost 3,400, Franklin more than 5,000;[48] the Confederate wounded, however, far exceeded the usual proportion to the killed. Subjected to a tremendous artillery fire, and hidden beneath dense timber, while many were injured by splinters of trees, few were mortally hurt. Taliaferro had but five killed out of a total loss of 172, and D. H. Hill about the same number. One thousand Federals were taken prisoners.

Shortly after three o'clock, a Federal brigade, with the purpose probably of covering by a diversion the movements on the left, debouched suddenly from the Deep Run ravine, turned the left of the line of scouts, captured a party of fifteen men, and, deploying left, advanced against the right of Hood. Unsupported, and disconcerted by a sudden counter-attack of two regiments of conscripts, delivered with that impetuosity for

48. [This figure obviously includes the casualties of the two divisions of Sickles and Birney, on loan from the Third Army Corps (*Official Records*, XXI, 133–42).]

which Hood and his division were so famous, it was easily repulsed. The pursuit, continued for some distance, was wisely stopped by Hood, though to the chagrin of the soldiers, before it came within reach of the enemy's main line.

Right Attack.

[2 p.m.] Shortly after two o'clock, about the same moment that Meade's division began to fall back before Jackson, Sumner ordered Sturgis's division of the Ninth Corps to make a last bid for victory. The Fifth Corps, belonging to Hooker's command, the Centre Grand Division, held back hitherto in reserve beyond the river, was at this time passing over the bridges; and Griffin's division,[49] which had already crossed, was sent in to support Sturgis. Under cover of a heavy artillery fire, to which Lee's guns did not reply, and the musketry of the broken Second Corps, which still clung to the ravine of the mill-sluice, Sturgis moved forward from beyond the mouth of Hazel Run, and Griffin from the town.

As the Federals advanced, General Ransom, commanding on Marye's Hill, having observed the passage of the Fifth Corps across the bridges, and probably anticipating a departure from the feeble tactics hitherto pursued, brought up two regiments (of Cooke's brigade) from the second line on to the plateau, and moved forward two, which had been stationed in front of the house, to the crest of the slope.

This assault, though made on a broader front, fared no better than those which had preceded it. Sturgis lost 1,028, and Griffin 818. One or more battalions reached the shelter of a hillock 150 yards in front of the right shoulder of Marye's Hill, and the Confederates were unable to drive them from their position. Towards dusk they were reinforced by a battery.

The Fifth Corps, having crossed the stream, took post in the following order (relieving respectively Hancock, French, and Howard): Griffin on the right, Sykes in the centre, and Humphreys on the left leaning on Hazel Run. The remnant of the Second Corps now took refuge in the buildings of the town and beneath the bluffs which stand above the river-beach. Whipple's division still retained its first position.

Two corps d'armée broken and demoralized, an enemy out-

---

49. Sturgis, 3,500; Griffin, 7,500.

numbering his own unshattered troops,[50] invigorated by success and holding a formidable position, were the conditions that now faced Burnside; and it ought to have been plain, even to him, that with his dispirited army his only course was to abstain from further assault, and either to withdraw across the river,

50. So, according to his own estimate of Lee's force, he must have believed.

SKETCH II

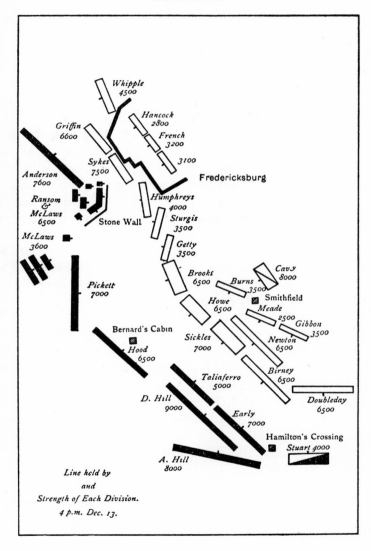

The Campaign of Fredericksburg

or to strengthen the line of defence he already held, and make every effort to maintain it until reinforcements should arrive. Franklin could not advance; Marye's Hill was impregnable; to retreat or to remain were alike hazardous, and every man and gun would be needed to ward off disaster. Everything pointed out that further assault would be worse than useless; but even the bloody experience of the day had not taught him wisdom. Wielding, like an angry child, the authority which a government, regardless of the lessons of history, and forgetting that men's lives are the price of a general's incompetency, had entrusted him with, he sent in a fresh corps d'armée, unsupported by any movement of the Left Grand Division, to dash itself to pieces against that invulnerable hill.

It is true that at 2.30 p.m. he had ordered Franklin to make another forward movement, but, as we know, that general, if not unable was at least unwilling to obey. The fresh attack on Longstreet did not take place till 4 p.m., and at that time Burnside must have been well aware that the left was in no condition to make a simultaneous attempt.

Unmolested by Franklin, General Lee was free to turn his whole attention to Sumner and Hooker, to employ all his resources, if necessary, to overwhelm them. He could have had but little fear of this last effort of his baffled and desperate foe.

[4 p.m.] It was not, however, without remonstrance on the part of General Hooker that further lives were sacrificed. To him it was apparent that the army had lost heart, and, riding over the river to headquarters, he endeavoured to dissuade his superior from his mad resolve. Counsel and expostulation, however, were alike unheeded, and, as he returned, Getty's division of the Ninth Corps, advancing from below the mouth of Hazel Run, attempted to turn the shoulder of Marye's Hill. As the attack developed, Kershaw brought up the 15th South Carolina and 16th Georgia, and with them stiffened the threatened point. The 15th South Carolina was "trebled" on the Phillip's Legion, and the 16th prolonged the line to the right. However, the Confederate infantry had but little share in Getty's repulse.

The Federals, advancing obliquely to the line of fire from Lee's Hill, were enfiladed (as Sturgis's troops had been) by two of the batteries posted thereon; crossing the unfinished railroad embankment they were thrown into confusion, and no sooner did they reach the open and feel the musketry than they shrank back to the nearest cover, and made no further effort to storm the breastwork.

At the moment Getty moved out of the ravine of Hazel Run, Walton's guns, occupying the pits on Marye's Hill, their ammunition being expended, were in the act of giving place to two of Alexander's batteries. The relief was delayed by the overturning of the leading gun on the steep and narrow track, and did not come into action. To this fact we must in part attribute the smallness of Getty's loss, which amounted only to 207 officers and men.

[5 p.m.] It was now the turn of the Fifth Corps. Twilight was falling fast, and the incessant flash of musket and cannon shone redly through the gloom when the last devoted Federal division advanced to the assault. In this final attack, however, more method was pursued. Hooker had made, as far as possible, a careful reconnaissance of the position; and, after consultation with the corps and divisional commanders, had determined to advance with unloaded rifles and to rely upon the bayonet alone.

The attack was entrusted to the two brigades of Humphreys' division. Sykes's division of regulars followed in close support in echelon upon the right. To cover the advance two batteries were stationed on the left of the Telegraph road, within 440 yards of the hill, and four to the right rear. Four guns, under cover of the gathering darkness, were actually brought up to within 150 yards of the stone wall; but there, having revealed their presence by a volley of canister, were overwhelmed by so fierce a storm of musketry that they were immediately abandoned.

Allabach's brigade led the advance. While they were still more than 100 paces distant from the hill, the soldiers, unused to yield implicit obedience to those who led them, took upon themselves the conduct of the fight, halted, and began to use their rifles. Exposed to the full force of the Confederate fire, they were unable to hold their forward position and quickly gave way.

Then Tyler's regiments, bracing themselves for a more determined effort, charged forward swiftly and silently through the darkness. But where so many had failed how could they hope to succeed? The pitiless round shot, smiting them front and flank, dashed them on the corpse-encumbered field. The hill, which rose through the smoke a shadowy mass before them, streamed and throbbed with flame, and the pitiless musketry piled its victims high within the limits of the Slaughter-pen. There, threescore paces from the wall, the line faltered; bewildered by the carnage, yet disdaining to fly, the men began to load. In vain the gallant Humphreys urged them on; those that were left unhurt could do no more; and after a brief and irregular fire, the relics of these fine regiments fell slowly back to the ravine. So complete and speedy was the repulse, that Sykes's division was utilized only to cover the retreat. Humphreys lost 1,760, little less than half his number.[51]

Shortly after this attack, General Longstreet ordered up Kemper's brigade of Pickett's division to Marye's Hill. He appears to have anticipated a bayonet charge under cover of the night, little dreaming that such an attempt had been already made. At the same time, the regiment holding the trench between the plank and Telegraph roads, which had lost heavily and had expended its ammunition, was relieved by a battalion from the second line. There was no need, however, of further precaution against attack. Humphreys' charge was the final effort of the Federals, and with his repulse the battle ceased.

During this last assault Marye's Hill had been held by seventeen battalions, about 7,400 men. Although nearly 30,000 Federals had been arrayed against it during the day, and it had been exposed to the fire of the heavy batteries above Fredericksburg and near Falmouth, the loss of Longstreet's corps did not exceed 1,550; while that of the enemy in this quarter of the field alone amounted to more than 7,000, one-fourth the strength of the force actually engaged. It was thanks only to the forbearance of General Lee, who restrained his artillery from overwhelming the town, filled with stragglers, wounded, and demoralized troops, that the casualties of the Union army were not far more numerous.

To the vicious tactics adopted by the Federals the disastrous result was in great measure due. The soldiers, though they went

---

51. [According to the official casualty return, Humphreys lost 1,019 (*Official Records*, XXI, 137).]

into battle without confidence, fought with admirable determination, and it was through no fault of theirs that success escaped them.

In the first place, notwithstanding the large force of heavy guns available, little or no preparation was made for the attack; and the very obvious rule that "the stronger the enemy's entrenchments, the longer the preparation, and the greater the quantity of artillery fire required," was either unrecognized or neglected.

Secondly, the assaults were isolated and unsupported, and no attempt was made to distract the enemy's attention by demonstration right and left.

Thirdly, the formation in which the troops attacked was most unsuitable. On a front, probably never exceeding 600 yards in length, their lines of from 1,200 to 2,000 men advanced at several minutes' interval against a strong position held by a large force of veteran riflemen and numerous batteries. The very formation, occupying as it did but a limited front, compelled the enemy to concentrate his fire: and so wide was the distance between brigades that they were swept away in detail. No attempt was made to reinforce the leading brigade as it reached charging distance; the wave of attack was not propelled forward by fresh impulse from the rear, but left to rely upon its own momentum and to expend its force ere it reached the mark. In their earlier attempt on Plevna, the Russians, it will be remembered, committed this last error with like fatal result.

Again, it showed little tactical judgment in the leaders that the attacking lines were allowed to halt and fire. Every moment the men remained halted but added to the death roll and produced no effect whatever upon the well-protected enemy. The only possible chance of storming the wall, under the conditions adopted, was by a swift, unfaltering rush, which might have carried a few of the assailants within the line of the defence; where, promptly supported, they would have made good their footing and opened the way for a successful advance.

Lastly, the force employed at any one period was wholly insufficient, not only against Marye's Hill but also against the Confederate right. Including Birney's division, 22,000 infantry, one-fifth of the total strength, was the utmost number employed at the same time in the assault, and Birney can scarcely be said to have formed part of the fighting line. At 1.15 p.m., the hour alluded to, Meade, Gibbon (and Birney), on the left, and Howard on the right, advanced almost simultaneously. This was the only combined movement throughout the day. Burnside has been condemned for the fault of making a direct attack upon an unbroken and well-posted enemy, but the error, says Colonel Chesney, was older than his day; for Napoleon threw away the same number of men in a vain assault on the Russian entrenched camp at Heilsberg, 1807; and it may be added that at Waterloo the French Emperor neglected, as the Federal general, to push local successes and to adequately support his attacking columns. Grant's battle of Cold Harbour in 1864, where 13,000 Federals fell in little more than ten minutes, has been included by Chesney in the same category as Fredericksburg and Heilsberg. But Burnside was guilty of a far graver error than was committed at Heilsberg, Waterloo, or Cold Harbour. There, whatever other mistakes were made, the attacks were at least made in force. The following principles have been laid down by well-known military writers: "You can never be too strong when making an attack, for you can never be perfectly sure of what force you may encounter, or at what moment the defender may make a counter-attack." Again, "An attack on an enemy presupposes a superiority of force at the place where the attack is made. War is but the art of being the strongest at the right place at the right time. . . . for an attack to have a reasonable hope of success, the attackers at the point the attack takes place must be superior." Burnside scarcely appears to have grasped these important principles, and his method of feebly "tapping" (as has been happily said) at different points of the enemy's line is a blunder of which there are few instances. He had, it is true, to provide for the security of the bridges in case of repulse; but it may be observed that a vigorous attack is the very best means of ensuring unmolested retreat, especially when engaging a smaller force. Though the assault should fail, the enemy will probably have suffered too severely to pursue with energy. Wellington has been condemned for failing to follow up his victories, but on more than one occasion his troops were so reduced and exhausted by sheer hard fighting against superior numbers, that they were physically unable to reap the fruits of success.

A minor detail connected with the operations against Marye's Hill may here be noticed. If the attack had succeeded and the wall been carried by the division forming the fighting line, it would have been necessary to send forward supports from the town: for their speedy transmission to the scene of action, provision was neither made nor contemplated. The mill-sluice was crossed by two wooden bridges, one of which had been partially destroyed. No attempt was made to repair it during the battle; the advance, therefore, of reinforcements would have been long delayed; and the lengthy column exposed on the open road during the slow passage of the obstacle by the leading troops would have afforded a fair mark for the enemy's artillery.

The practice of the Federal guns, though in every respect aided by position, must have been far from accurate; one gun only was disabled on Marye's Hill, none on Prospect, and the casualties among the Confederate gunners were but trifling. Walton lost three men killed and twenty-three wounded; Alexander, one killed, ten wounded, and fifteen horses; Walker, on the right, three killed, twenty-four wounded, and ninety horses.

Movements of Jackson's Corps.

[5 p.m.] Shortly before dusk, seeing that the Federals did not intend to risk another attack, Jackson (doubtless with Lee's approbation) directed his divisional leaders to make ready for a general advance. All the available artillery was ordered to the front for the purpose of preparing and covering the movement. Stuart on the extreme right had already driven in Doubleday's skirmishers, and had pushed his guns forward as far as the Old Stage road. The divisional batteries were in position, and those which were to lead the way had advanced a hundred yards beyond the railroad, when so tremendous an outburst of fire was evoked from the guns on the Stafford Heights and behind the embankments of the road, that Jackson countermanded the order and, for the time at least, abandoned all idea of attack. In his report he stated that "the artillery of the enemy was so judiciously posted as to make an advance of our troops across the plain very hazardous." It would have been necessary to subdue the fire of Franklin's field guns before the Southern infantry could move out. Of that the hour did not permit.

During the night the sentries of the two armies stood within a hundred paces of each other.

The two following days saw the belligerents holding the same positions; but though there was much skirmishing and artillery fire, no offensive movement was undertaken by either. The Confederates, immediately the battle ceased on the 13th, set to work to strengthen their line by rifle-pits and abattis, in expectation of further assault. The Union troops were occupied in improving the natural defences of the Old Stage road; a series of entrenchments was constructed on either side of Hazel Run, and the streets of Fredericksburg were strongly barricaded.

Fortunately for the thousands of wounded Federals who lay during forty-eight hours helpless and unaided between the hostile lines—for Burnside, loath to acknowledge his defeat, did not until past noon on the 15th demand permission to remove them—the weather continued comparatively mild; but the time was one of frightful suffering to those who lay dying on the open fields, and of intense suspense to their comrades, expecting every moment to hear the thunder of Lee's guns and the terrible yell of his exultant infantry.

It will scarcely be believed that the force of folly could further go, and that the bloody lesson taught by the defenders of the stone wall should have been utterly lost; yet, on the night of the 13th, Burnside "as one of sense forlorn," announced to his generals in council that he intended, on the morrow, to attack Marye's Hill with the Ninth Corps in column of regiments. Providentially this scheme was overruled. For two days things remained *in statu quo*, and when on the night of the 15th a fierce storm of wind and rain burst from the south-west upon the valley of the Rappahannock, under cover of the tempest and the darkness the Federals withdrew across the river, and resumed their former position on the Stafford Heights.

The retirement was skilfully executed, does credit to the staff, and proves that the Northern army was by no means unwieldy.

The movement was kept a close secret up to the last moment,

and carried out in such carefully-guarded silence that the main body had already passed the bridges before the outpost line became aware that the operation was in progress.

Burnside himself wished to retain Fredericksburg and the bridge-heads, but was overruled by Hooker, who knew that there was a limit to Lee's patience, and that the fire of his numerous batteries being quickly concentrated upon the town and its occupants, the whole would be involved in the same destruction. It is said that the Confederates were already preparing incendiary shells for this very purpose.

The inaction of the Federal commander during the 14th and 15th we can easily understand. Fortune was kinder to him than he deserved. The hours passed away without his beholding the long line of gray-clad soldiery issue from the shadow of the woods, and sweep down the slope upon his demoralized masses; and, when he looked helplessly for an opportunity to escape, the very elements declared themselves on his side.

By almost every European soldier who has critically considered the battle of Fredericksburg, Lee has been censured for his supineness after the Federal repulse. If he had completed his work by the capture or annihilation of Burnside's army, the Confederate cause would have been well nigh won. Recognition by the European powers would in all probability have followed, the blockade have been raised, and the Union States, stunned by disaster, and realizing that the task of coercion was beyond their strength, have conceded the claims of their Southern sisters.

The chance of achieving this great result, of winning at a blow the prize for which he stood in arms, would assuredly have justified him in quitting his defensive attitude. One splendid success alone was needed; it was impossible by adhering to the defensive that such success could be won. What would Waterloo have been had Napoleon been allowed to retire unmolested?

In Lee's own words are given the reasons that he ventured no offensive movement: "The attack had been so easily repulsed and by so small a part of our army that it was not supposed that the enemy would limit his efforts to an attempt, which, in view of the magnitude of his preparations and the extent of his force, seemed comparatively insignificant . . . and we were necessarily ignorant of the extent of his loss." And again: "It was not deemed expedient to lose the advantages of our position, and expose the troops to the fire of his inaccessible batteries beyond the river by advancing against them."

These may be admitted as sufficient excuse for not delivering a counter-attack on the 13th, but they scarcely justify him in letting the two following days slip by. The very inaction of the enemy on the 14th revealed the extent of his discomfiture; and on the morning of the 15th, at latest, every Confederate gun should have been turned against the Federal lines. There was no time to be lost; the foe would either receive reinforcements or would escape; and, in either case, the Army of the Potomac would have to be encountered again on perhaps far less favourable terms.

That army was the finest body of troops beneath the Northern flag; it had been delivered into the hands of its enemy, and it ought never to have slipped from them to form, as it did, the solid nucleus of the force which not only saved the Union at Gettysburg, but finally, under General Grant, crushed the life out of the Confederacy.

Lee was influenced, say the Southern historians, by the thought of the suffering that would accrue to the inhabitants of Fredericksburg if he turned his guns, as he must have done in case of battle, on the already half-ruined town; and of the loss an offensive operation would have entailed upon his troops. Not for the only time in his life his judgment gave way to his humanity.

Soldiers were undoubtedly precious in the South, but, five months later, in conflict with this same Army of the Potomac, he lost at Chancellorsville 12,800 men and his great lieutenant, "Stonewall" Jackson. The destruction of Fredericksburg, and a far heavier loss of life than he had already incurred, would have been but a small price to pay for future peace and the triumph of the cause.[52]

52. [Henderson's criticism of Lee's inactivity after the battle has been answered by Sir Frederick Maurice, a prominent British military historian of a later generation. Maurice admits that Lee could have attacked on the evening of the thirteenth, had he realized that the Federal assaults had ended. But these "had been so easily repulsed" that more attacks were anticipated, "and to have left strong positions prematurely would have been foolish." In view of the fact that Burnside had reformed his men and was supported by a great mass of guns on Stafford Heights, Maurice contends that "to have attacked on the 14th would have been still more foolish" (*Robert E. Lee the Soldier*

The Federal position on the 14th and 15th was certainly strong, and was manned by a long array of batteries;[53] but it was commanded throughout by the Southern artillery, and exposed to oblique and even enfilade fire; for the heights beyond the Massaponax gave footing for long ranging guns.[54] Nor were the entrenchments along the Richmond road stronger than the successive breastworks carried so brilliantly at Chancellorsville by smaller numbers than were here available.

Moreover, the right [left?] flank, though the approach was obstructed by a narrow stream (see Maps), was open to attack. The assailants would doubtless have lost heavily from the enfilade fire of the batteries beyond the Rappahannock, but that the Federals at least did not believe such a manoeuvre impracticable, is proved by the disposition *en potence* of Doubleday's division.

On the right, the town of Fredericksburg and the bridges in rear would have been quickly rendered untenable and impassable, under the concentrated fire of Longstreet's guns, and the whole Federal army would have been driven to depend upon the narrow line of escape afforded by the three bridges near Deep Run. The best troops in the world would have been in danger of demoralization under such circumstances.

The heavy guns on the Stafford Heights would have been compelled to give their whole attention to the Confederate artillery, which commanded not only the town, the bridges, and the road, but also that space between Hazel Run and Smithfield,[55] where, driven from Fredericksburg, the Federal infantry would have been compelled to mass.

It must be borne in mind too that the fire of these batteries was plunging, little effective against troops moving in the open, and that the practice of the Federal artillery was by no means accurate. If at break of day on the 15th, Longstreet's and Anderson's guns, together with those of the reserve, had con-

---

[Boston and New York, 1925], p. 172). Freeman simply states: "No one who studies the ground can justly criticise him for failing to do so" (*R. E. Lee*, II, 466).]

53. Besides the thirty-six pieces which formed the *tête du pont*, Franklin had eighty field guns at his disposal.

54. The single Whitworth, posted here under command of Captain Hardy, is said to have done great execution on the 13th.

55. Smithfield is distant 2,300 yards from Prospect Hill, 2,500 from Bernard's Cabin, not beyond the range of the Confederate siege-pieces and rifled guns.

centrated their fire upon Fredericksburg; if Jackson, reinforced by one division of the First Corps, supported by Hood and Pickett on the left, and on the right flank by Stuart and two brigades of D. H. Hill's, had dashed against Franklin, while the uncertain light disconcerted the aim of the Federal gunners; if, at the same time, the batteries on Prospect Hill and near Bernard's Cabin, together with those beyond the Massaponax, had swept the ground in rear of the Stage road, the Northern army would in all probability have been driven into the waters of the Rappahannock, and, as Colonel Chesney has said, the awful scenes of Leipzic and the Beresina have been repeated in another hemisphere.

The long array of ordnance beyond the Rappahannock has been held by many writers as sufficient excuse for Lee's inaction; but neither weight of artillery nor superior numbers ever intimidated the Virginian general; and with the affair of Boteler's Ford still fresh in their memories, it is unlikely that his soldiers would have hesitated to fall upon the Federals, however numerous and commanding were the batteries that opposed them. The following spirited account of that action, which was fought a few days after the Antietam, is given in Dr. Dabney's "Life of Stonewall Jackson":—

"On the north bank of the Potomac were planted seventy pieces of heavy artillery, while, under their protection, a considerable force of infantry had passed to the southern side, and were drawn up in line upon the high banks next the river. Under the direction of General Jackson, Hill (A. P.) formed his gallant division in two lines and advanced to the attack, regardless of the terrible storm of projectiles from the batteries beyond the river. The enemy attempted for a time to resist him, by bearing heavily against his left, but his second line, marching by the left flank, discovered itself from behind the first, and advanced to its support; when the two charging simultaneously and converging towards the mass of Federals swept them down the hill and drove them into the river. The troops of Hill rushed down the declivity and, regardless of the plunging shot and shell of the opposing batteries, hurled their adversaries by hundreds into the water, and as they endeavoured to struggle across, picked them off with unerring aim. In this combat, General A. P. Hill did not employ a single piece of artillery, but relied upon the bayonet alone. Early was at hand to support him, but no occasion arose for his assistance. The whole loss of the Confederates was 30 killed and wounded. The Federals admitted a loss of 3,000 killed and drowned, and 200 prisoners."

It must at the same time be admitted that, if Lee in his turn had become the assailant, his loss would have been heavy.

The Federal line of defence was strong; and herein lay the great fault of the Confederate position. It was well suited for a force acting on the defensive absolute, but not for a force whose object was, besides holding its ground, to inflict a crushing defeat.

"Defensive actions," says Colonel Schaw, R.E., "may be classed under two heads—

"(a) *Passive* or delaying, in which the object is chiefly to ward off a blow, either to maintain possession of the ground or to gain time.

"(b) *Active*, in which the object is the defeat of the enemy, the defender fighting in a chosen position, with the intention of attacking his adversary when the favourable moment arrives."

That Lee intended to confine himself to the former is improbable. The passive defence was uncongenial to his bold and enterprising spirit, ill-suited to the circumstances, and there is no doubt that he was much disappointed at the retreat of the Federals on the night of the 15th. His error was in allowing that day to pass without attack. The fact remains, nevertheless, that the Fredericksburg position was faulty in this respect, that it was difficult for counter-attack. Stonewall Jackson, a soldier whose genius for war has been universally recognized, is said to have remarked some days before the battle, "I am opposed to fighting here. We shall whip the enemy, but gain no fruits of victory. I have advised the line of the North Anna, but have been overruled." Lee's original intention also had been to retire to the Annas, and to thus draw Burnside to a distance from his base. "My purpose was changed," he wrote to the President, "not from any advantage in this position, but from an unwillingness to open more of our country to depredation than possible, and also with a view of collecting such forage and provisions as could be obtained in the Rappahannock Valley."

The truth probably is that he believed the Army of Northern Virginia capable not only of holding the Fredericksburg heights, but also of assuming the offensive successfully, despite the embankment of the Stage road and the commanding Stafford Heights; and that it was the elements alone which robbed him of a decisive victory.

We know, however, from his own report, that the difficulty of a counter-attack was one of the reasons which held him back on the evening of the 13th and the two following days, and this lesson may be learnt from his inaction: in the selection of a field whereon it is intended to fight an active defensive engagement, it is above all things important that "no good positions to resist counter-attack," should be available for the enemy.

Curiously enough, a pamphlet, written by General Franklin in his own defence, reveals how very nearly Lee's expectation of further assault was realized, and how little his judgment was at fault after all. The Federal general states that at the council of war held after the battle, both Burnside and himself were in favour of renewing the attack on the next day, and that it was only after long discussion that they were overruled. Prudence prevailed also in the Confederate councils; the Federals were allowed to glide from the toils; and five months later, the Army of the Potomac, more numerous and as well-equipped as ever, once more stood face to face with its old opponent. The scars of Fredericksburg had quickly healed, and the Union Government had gained more than it had lost. At last the President and his advisers had abandoned the idea that a general "nascitur non fit;" had recognized that the art of war is something of a mystery after all, and that to make an efficient leader, study and experience must be added to "a cool head and a stout heart."[56] An untried man never again commanded the Army of the Potomac.

It may be interesting to compare the battle with a more recent and greater conflict, and by that light to consider the tactical combinations of the American generals.

The position of the French army at Gravelotte answers in one point at least to that of the Confederates at Fredericksburg,

56. *Vide* Lord Macaulay's Essay on Hampden. [*Critical and Historical Essays* (London, various editions), Vol. I, chap iii.]

and the numbers of the opposing forces bore almost the same proportion to each other as those of the North and South.

Both positions were naturally strong; of both the weak flank, strategically and tactically, was the right. The force employed against this point by the Germans was 140,000 infantry; by the Federals 50,000; eight divisions in either case. The German frontal attack was carried out by five divisions, all of which were vigorously engaged; one was held back in reserve, while two, marching round the right flank of the enemy, eventually delivered the decisive blow.

Franklin employed four divisions in the first line, of which two only were vigorously engaged; two were merely a containing force, and did no more than skirmish: two were in second line; one, together with the cavalry, constituted the reserve, and one, *en potence*[57] on the left, was held in check by two brigades of cavalry.[58]

With the enormous artillery force at his disposal, and a strong defensive line in rear of his attack, Franklin need not have retained more than one division at most to cover the bridges. A portion of the Federal cavalry might have relieved Doubleday. Five divisions would have been then available for offence. It will be remembered that besides De Russy's sixty-one guns upon the Stafford Heights, a great battery of thirty-six pieces covered the bridges. The German writer before quoted remarks that, "the reserve which infantry proceeding to the attack leaves behind should be reduced to a minimum, being only intended to cover its rear in case of need, perhaps to hold a defile which may happen to be in dangerous proximity; it will be best if this reserve for the infantry is furnished by the other arms." Franklin, it cannot be doubted, was far too solicitous for the safety of the bridges; and his dispositions were made with a view of securing his own retreat rather than of compelling that of the enemy.

57. Speaking of this formation Jomini remarks:—"A *crochet*, or as it is better known, the order *en potence*, if used to protect a flank against an enemy that can manoeuvre, is a remedy worse than the disease it is used to cure." Against the long-ranging fire of the present day it would be impossible to retain troops in this formation: a flank *en l'air* must be protected by troops placed in echelon.

58. In first line, Meade, Gibbon, Howe, Brooks; in second line, Birney, Sickles; in reserve, Newton and the cavalry; *en potence*, Doubleday.

As regards the conduct of the fight on the right wing, the same error seems to have been committed by both the Germans and the Federals.

The action against the French left is thus criticized by General Hamley, and his words apply equally well to the useless attacks on Marye's Hill: "Where circumstances would already seem to indicate that the real effort must be made elsewhere, troops are sent on in an indiscriminate fashion. Thus the German right was separated by a ravine from the French left; it rested on the Moselle, and the French could alone advance on that bank by forcing the Germans back. The decisive attack was to be made by the German left; the ground was excellent for defence, bad for offence; all good reasons for the avoidance of unnecessary slaughter and the risk of defeat, by slightly engaging or even refusing that wing of the German army. Yet the corps there were hurried to the attack no less earnestly than at other parts of the position, suffering enormous losses . . . and inflicting only the most trifling damage on the opposing corps."

The first repulse of the Guard Corps before St. Privat was due to want of sufficient preparation, as was Meade's first repulse at Fredericksburg. To the measure of success afterwards achieved the co-operation of the artillery in both cases largely contributed. The Federals, however, though the bombardment of Jackson's position was heavy and prolonged, failed to adopt the expedient of concentrated fire. The numerous batteries acted independently of each other, and not as one command. This fault, however, has been conspicuous in more highly trained armies than that of Burnside. The German artillery in '66, the French in '70, and the Russians in '77, followed the same bad tactics.

Perhaps the most essential difference in the conduct of these two great battles is the decisive influence exercised by the superior and staff officers, for good in the one case, for evil in the other. The intelligence and energy displayed by the German corps, division, and brigade leaders in working out the tactical end of the Commander-in-chief, stand out in bright contrast to the feeble and ill-concerted tactics of the Federal commanders of like rank. It is not too much to say that Germany owed her success in 1870 to the sound practical and theoretical training of her officers, and that to the want of such training the disasters of the American armies were mainly due.

If we compare the defensive dispositions of Bazaine with those of Lee, we shall find that a most important difference exists. The Confederate general reserve, D. H. Hill's division, was placed in rear of the weaker right flank. Not so the French Guards, but in rear of the stronger left, seven miles distant from the spot where the battle was decided. "What might have been the result," says a military writer, "had the French Guards been on Bazaine's right flank, and been able to assume the offensive when the Prussian Guards' attack was first brought to a standstill?"

Lastly, the fighting at Fredericksburg furnishes another proof of the truth of that rule so terribly demonstrated at Gravelotte, that good infantry, sufficiently covered, and with free play for the rifle, is, if unshaken by artillery and attacked in front alone, absolutely invincible.

Six frontal attacks failed against Marye's Hill. On the left, Meade's division won a measure of success, because his leading battalions, having penetrated Hill's first line, were enabled to take the opposing troops in flank. Archer's brigade was partially routed, and would have been wholly so, if the flank attack from the thicket had been vigorously followed up. Lane was forced to give ground, Gregg's conscripts were driven from the field, and if the Federal brigadiers and battalion leaders had been more intent on developing local flank attacks at the same time that they pushed the enemy in front, their success would have been far greater.

Nothing destroys the morale of men in action so speedily and effectually as a flank attack, and except by this method good infantry will seldom be beaten. Such, it has been said, are now "the difficulties and dangers of mere frontal attacks," that "every assailant will in all probability attempt to turn one or both flanks of the enemy's position." It is the main idea of modern tactics. All officers, however small their command, should bear this in mind, and seize every opportunity of bringing a flanking fire to bear upon the enemy and of developing local flank attacks. If, for example, a body of troops break in at some weak point of the defence, a portion, let them be ever so few, should at once be directed to face right and left and enfilade the enemy's line. The efficacy of such a fire may be gathered from the following instance, witnessed and recorded by General Valentine Baker: "Our two advanced companies now opened a heavy enfilading fire upon the left of the Russian line.... The effect was most extraordinary. The whole left wavered for a moment, and fell back in confusion.... Then the panic extended all along the line, and it retreated precipitately." How fully General Lee recognized the truth of this principle is proved by the fact that each of his offensive victories was won by a terrible stroke on his adversary's flank, delivered by his great lieutenant, Stonewall Jackson.

As to the failure of the Federals against Jackson's division, it must be remembered that the troops were engaged in thick covert; and of all species of combat, wood-fighting is the most difficult, demanding for successful prosecution the highest training on the part of the company officers and men.[59] In few situations has the action of the subalterns and section leaders so much influence upon the issue of the battle, for here, writes Colonel Home, "the direction of events must, to a great extent, pass out of the hands of superior into the hands of subordinate officers. Consequently, self-confidence and habits of individual action within certain limits, are, for such fighting, invaluable." Now, such habits, invaluable under all circumstances, may be attained and fostered by study and thought; and an officer, thoroughly versed in the theory of tactics and accustomed to work out practical problems on the map, will, if he is cool and resolute, be able on active service to turn his knowledge to account, to recognize and profit by the opportunity.

The Germans, in the war of 1870–71, proved far superior to the French in wood-fighting; and "the cause," says Colonel Home, "which produced this result, was the Prussian method of instruction, which had developed the individual force of the soldiers, and had accustomed many subordinate officers to the responsibility of command."[60] That is, the German soldiers

59. May, 1864, eighteen months later, saw this same Army of the Potomac battling in the dense forest of the Virginian wilderness. Throughout that desperate campaign the troops were handled by officers of all ranks with rare skill and judgment. The lessons of Fredericksburg and Chancellorsville had not been wasted.

60. The group system, so strongly advocated by those who propose a reform of our infantry drill and formations, would doubtless bring about the same results, both in men and officers.

owed their efficiency in this kind of fighting (and much of it in all others) to the thorough training of the company officers and the men they led. The eminent military critic, Von Boguslawski, thus compares the training of the Prussians and French: "With us every officer gains a knowledge of the tactics of the three arms during his carefully-conducted theoretical studies; we have also made use of our previous experience in war. The French were wanting in both ways. Their line officers undergo no scientific examination in military matters, . . . and, lastly, it was no business of the officers generally to reflect upon former experience."

The theoretical studies of our volunteer officers must perforce be "personally conducted," but let them be assured that if they have drawn on the experience of others, if the golden rules of tactics are, as has been said, tattooed upon the brain, in the hour of action they will find themselves instinctively doing the right thing.

Much instruction may be gathered from a consideration of Lee's tactics and the disposition of his troops.

His defensive measures could scarcely have been improved upon. The judicious manner in which his supports and reserves were placed, and his line so manned as to leave little more than a third of his force exposed, yet completely covering his front, cannot be too highly praised.

The following points are well worthy of remark:

The first line (except at one point) contained as many guns and rifles as could be usefully employed thereon.

The second, of equal strength, was disposed at convenient distance, prepared, as soon as the real attack developed itself, to move up rapidly to sustain the first, or to drive back an intruding force. This line was drawn from the same division that supplied the first; thus the fighting line and supports were, at all points, under one command.

The reserves were not massed in one locality, but distributed at various points: they belonged to the same corps or division as the troops in front of them, and there was, therefore, unity of command throughout the whole depth of the position.

In rear of the right—the weak flank, and at the same time that from which a counter-attack must have been delivered—was posted a strong, compact division. Here, too, numerous batteries were held in readiness to oppose any turning movement up the valley of the Massaponax, or, in case of counterattack, to reinforce the artillery (under Pelham) to the right front.

The strongest sections of this position, viz., the ridge N. of Marye's Hill and the central re-entrant, were but thinly occupied.

The action of his artillery was perfectly adapted to the circumstances. Unable to cope with the more powerful and commanding ordnance of the Federals, the batteries reserved their fire for the advancing infantry. So well were the positions selected, and so good was the practice, that more than once the enemy's fighting-line was driven back by the guns alone; and Gibbon's division was, at the last assault, brought to a standstill and utterly demoralized by the thirty-three pieces which confronted it. Again, the guns were massed as far as possible, not distributed among the infantry divisions; and the combined action of these large batteries appears to have been most effective. Pelham's gallant and remarkable intervention, early in the day, needs no comment.

The "incomparable Confederate infantry," excellently handled throughout, showed itself, as ever, stubborn in defence and impetuous in attack. The ceaseless storm of shot and shell which crashed with terrific uproar through the leafless woods upon the right affected its courage not a whit; and the defenders of Marye's Hill, though exposed to repeated and desperate assaults, maintained their sangfroid and steadiness to the end.

Yet at this point, although Sumner and Hooker were always easily repulsed, the front line of the defence lost somewhat heavily. Kershaw's brigade,[61] actively engaged upon the hill during the afternoon, lost 373 out of 2,400; and one regiment, the 3rd South Carolina, lost six successive commanding officers. The regiments actually behind the wall suffered in the same proportion. Cobb's brigade, which held that position throughout the day, lost 234 out of a total strength of 1,600.

Longstreet, on the left centre, adhered strictly to fire tactics; and, though the opportunities were tempting, the defenders of

---

61. This brigade fired an average of 55 rounds per man. Walton's nine guns expended 2,400 rounds; Maurin's battery of four pieces, to the left of the Orange road, 200; and that of Wilcox [Lewis' Light Artillery] covering the ford, 400.

the stone wall made no attempt to dash out with the bayonet on Sumner's or Hooker's shattered ranks. The corps commander depended on fire alone to clear his front; and no useless counter-attacks were permitted to swell the death-roll of his divisions. It may be here noticed that, before the battle commenced, he had instructed Hood, if opportunity offered, to attack in flank any force which should assail the right wing. Unfortunately that general did not consider that such action was at any time expedient. The manner in which reinforcements were brought up to the stone wall as each column of attack advanced, thus giving *moral* support to the first line of defence and ensuring a sufficiency of ammunition, was most judicious. All such orders emanated from the corps commander.

Atkinson and Hoke certainly acted with much indiscretion in pursuing Meade's routed division across the plain. Such partial and local counter-attacks, when once an intruding enemy has been driven from within the position and deprived of his rallying point, are dangerous and useless. It is wiser to leave the retiring lines to the unobstructed fire of artillery and musketry, and not to expose isolated brigades or battalions to inevitable loss by permitting them to encounter, unsupported, the enemy's reserves. However, this is a fault which all high-mettled troops are apt to be led into; and the instance of the rash advance of the Guards and German Legion at Talavera will occur to all readers of Napier. "In the excitement of success," says that historian, "the English Guards followed with reckless ardour, but the French reserves of infantry and dragoons advanced, the repulsed men faced about, the batteries smote the Guards in flank and front so heavily that they drew back, and, at the same time the Germans being sadly pressed, got into confusion."

Wellington was far, however, from condoning such imprudence, and his letter on this subject, dated May 15, 1811, ought to be carefully studied and taken to heart by every officer.

The arrangements for renewing the ammunition-supply during the fight were evidently defective. A regiment, once engaged, had no means of replenishing cartridge-boxes, however great the need, except by retiring *en masse* to the wagons. Thus regiments which had exhausted their rounds became suddenly ineffective at the very crisis of the fight, and were compelled to withdraw or give place to others. The dread of such a contingency doubtless made the soldiers careful, and prevented waste; but the want of an adequate system was liable, in the heat of an engagement, or when troops had got out of hand, to lead to disaster and defeat.

Lane's brigade, for instance, gave ground before Meade's attack from sheer inability to reply to the enemy's fire. Atkinson, brought to a stand by Birney, lost heavily for the same reason; and the clumsy four and even six-deep formation employed along the stone wall was dictated rather by fear of ammunition failing than of the enemy's onset. Lee was so fully alive to this difficulty and to the danger of rapid firing that, seeing no means by which they might be overcome, he was averse to the employment of breech-loading rifles. "What we want," he remarked, "is a fire-arm that cannot be loaded without a certain loss of time; so that a man learns to appreciate the importance of his fire, and never fires without being sure of obtaining a result."[62] This opinion of a great and experienced soldier, *speaking of troops whose standard of discipline was not a high one,* is worth consideration.

No doubt the lack of discipline and the inferior training of the officers increased the difficulty. Company and section leaders cannot reflect too seriously upon this point: their responsibility is great, for, practically speaking, they alone can control the expenditure of cartridges. It is their bounden duty not only to prevent waste in the excitement of action, but so to train their men in peace that the soldier will consider every round as an article of price, not to be parted with unless he is "sure of obtaining a result." Volley-firing, and limiting the range against infantry to five hundred yards at most, are the surest means of providing against the want of ammunition at the supreme moment; and the sooner it is recognized that long range fire is

62. [It is curious to note that never once in all his writings does Henderson mention Major Justus Scheibert, who certainly wrote more about the Civil War than any other contemporary foreign military observer save perhaps Lord Wolseley. Yet this passage would suggest that he was familiar, either directly or indirectly, with the opinions of the official Prussian observer, for it was to Scheibert that Lee expressed—in words of which Henderson's quotation is a direct translation—his misgivings about the breech-loader (Scheibert, *Der Bürgerkrieg* [Berlin, 1874], p. 27). There are other passages in Henderson's works that seem to reflect views expressed earlier by Scheibert.]

a special weapon, to be used only on special occasions, the better for the efficiency of our infantry in general.

It is ordered that the system of renewing the supply be practised on the drill-ground; still men must not be trained to believe that wherever they are the ammunition-cart will invariably be at hand;[63] they must be taught to husband the supply they have about them, and to rely on that alone.

A strong Federal division was neutralized throughout the battle by Stuart's dismounted troopers; and it was decidedly advantageous to the Confederates that they were enabled thus to utilize this arm, precluded as it was from exercising its proper functions.

It is curious that the Confederate staff did not recognize the importance of the projecting tongue of the forest so often alluded to. Even had it been impassable as was supposed, it covered the advancing enemy from artillery fire, and gave him a place of shelter where he could mass his troops in security. It was, in fact, a bastion, insufficiently provided with flank defence, and the head ought in any case to have been held by a line of rifles, and the occupation of the dead ground in front provided against.

The want of vigilance which permitted the Federal escape is inexcusable. Such carelessness in the performance of this most important duty is the inevitable outcome of loose discipline and indifferent training. The wearisome and monotonous work of the outpost line is of all phases of field service the most distasteful, especially to volunteer soldiers, who are unused to the exactions of discipline, and have not attained the habit of constant and exact fidelity to duty.

Burnside's escape shows clearly and forcibly the evils that may arise if the precaution of patrolling is neglected, and from the general slackness apt to prevail after a successful engagement. According to General Hooker's report, it was after eight o'clock in the morning when his last troops left the town, a statement which shows up the lack of enterprise and watchfulness of the Confederate picquets in a most glaring light. Nor

---

63. The Turkish practice of carrying the regimental reserve on pack-animals is far superior to our own system, for mules or ponies can travel over ground impassable for carts.

can General Lee be acquitted of responsibility. There was a possibility that his adversary might slip away. The night was dark and stormy, and he himself, after the Antietam, three months before, under somewhat similar circumstances, had eluded the Federals in this very manner. Special orders ought to have been given as to constant and vigorous patrolling, and scouts have been instructed to penetrate at all risks the enemy's lines.

Want of vigilance, too, we may notice, appears to have been the cause of the disaster on Archer's left. The wood on that flank was deemed impenetrable, and left unoccupied and unwatched; and owing to this neglect of the rule that all ground to front or flank must be under observation, the brigade was surprised and partially routed. *How often in military history has success been achieved by a movement over ground deemed impassable, and therefore left unwatched?*

Moreover, had scouts or connecting links been posted between the Confederate brigades in first line and their supports upon the road in rear, Gregg's brigade would have then received warning of the Federal advance, and have been prepared to resist it. The neglect of such small precautions often gives great opportunities to the enemy; and it is in attention to such details that an officer shows his efficiency.

The large artillery force held in reserve by General Lee calls for some remark. Very nearly one hundred guns never came into action. Since the Franco-German War the practice of holding back guns in reserve has been generally condemned. Colonel Home has laid down that "to place guns in reserve is to voluntarily deprive one's-self of a most powerful auxiliary at the very moment its aid is most urgently required. And, further, it would seem that there is no reason why all guns should not be brought into action as soon as possible, and remain in action until the close of the engagement." Lee had, however, excellent reasons for not employing all his batteries. The first was scarcity of ammunition. On the night after the battle there remained barely enough for one more day of conflict, and it was not until the evening of the 14th that fresh supplies came up from Richmond. Secondly, all positions available for artillery on the right were fully occupied. On the open ridge between Lee's Hill and the river it would have been useless to post guns except under cover, and the emplacements provided were already tenanted. If he had attacked on the 15th, there is no doubt but that every single battery would have been utilized, however great the risk. At the same time it may be questioned whether, in view of a possible counter-attack, it would not have been well to have brought up into front line the twelve guns held in reserve behind Hood's and Pickett's infantry. The Federal Sixth Corps, immediately opposed to these divisions, suffered severely, as it was, from artillery fire. If it had been assailed throughout the day by a larger mass of guns, it would have been much shaken, and perhaps in no position to resist the onset of two of the finest divisions in the Confederate army.

Unfortunately for his cause and for his military reputation, Lee confined himself strictly to the defensive. As must always happen in such circumstances, his success was but partial, and the glory he won was the only reward of victory.

Before we turn to a very brief consideration of Burnside's conduct of the battle, the following extract from General Sir P. MacDougall's "Campaigns of Hannibal," which is curiously applicable to Lee's tactics at Fredericksburg, may prove interesting. Besides affording us an opportunity of comparing the tactical skill of the great American soldier with that of the greater Carthaginian, it goes far to prove how invariable are the broad rules of the military art. For the Trebbia read the Rappahannock; for Roman, Federal; for Hannibal, Lee:—

"The confidence Hannibal felt in the superiority of his own genius is manifested by his plan for fighting at the Trebbia.

"In judging of ordinary men we should be inclined to censure the inactivity which permitted the Roman army to cross the river and form leisurely on the bank, without taking advantage of the confusion necessarily occasioned by such an operation, to attack and defeat it when landing, before it could recover from that confusion. A general less confident than Hannibal, when about to engage his troops against an untried enemy, would have availed himself of the most obvious and certain method of inflicting defeat. But Hannibal's policy was not only to defeat but to destroy, and by the moral effect of the annihilation to intimidate the Romans. . . . Had Hannibal attacked the Romans during this passage of the river, their defeat would have been less decisive, both in fact and in its moral effect. A much smaller number of Romans would have fallen; and both they and Hannibal's allies might have entertained, the one the boast, the other the reflection, that had the terrible Roman legions been arranged against Hannibal on a fair field, the result might have been very different."

So far the parallel, both of situation and of action, is singularly close; but turning to the last paragraph, we see in what essential respect the modern general fell short.

"When the Roman infantry came to close, their courage and discipline seemed capable of restoring the balance; but at the critical time Mago's ambush burst upon their rear, while the victorious Carthaginian cavalry charged both their flanks, and Hannibal pressed them in front. No troops could withstand such an onset. The centre legions indeed, overbearing all opposition, burst through their opponents and marched clear off the field; . . . but the remainder were driven back into the Trebbia with tremendous slaughter. Only a small remnant reached the opposite bank." The Carthaginian position was evidently chosen with the primary view of making the counter-attack, which was carried out with such decisive effect. Lee, on the other hand, neglected his opportunity, and "in war as in politics," said Napoleon, "the lost moment never returns."

A last word as to Burnside. In the preceding pages many grave faults have been laid to his charge, but yet we cannot condemn him as wholly wanting in capacity. A more enterprising commander would have certainly attempted to deal with the Confederate wings in detail before changing his base, but the Fredericksburg-Richmond line of operations, with a navigable river on the further flank, was, *under the circumstances*, a sound choice. After the arrival of pontoons and the accumulation of supplies he wasted valuable time, but he probably felt that, winter being close at hand, it was too late for action, and only yielded his better judgment to the irresponsible critics of the press. "Male imperatur cum regit vulgus duces," wrote Seneca, yet in these days of special correspondents and daily telegrams, the general, who, true to his own purpose, carries out a Fabian policy, must be a man of rare strength of mind. Delay will be denounced as failure; sympathy will give way to accusation, and the task be fourfold more difficult than even in the days which saw the occupation of Torres Vedras and the retreat from Burgos.

In withholding Sumner from crossing the river and seizing Fredericksburg and the heights beyond when the advanced Grand Division first reached Falmouth, he acted with due discretion. The town, it is true, was held only by an insignificant force, but the only means of communication were the fords; the river was rising rapidly; no pontoons were available; supplies were scarce, and Sumner's troops would have been exposed to attack in a position where they could not have been easily supported; and where, if defeated, they would have been destroyed. Moreover, it is doubtful if the whole Grand Division with its artillery would have had time to cross before the arrival of the Confederate First Corps.

Again, his arrangements for crossing the river and occupying the left bank were well conceived and skilfully executed, and compare favourably with Napoleon's imprudent conduct at Essling, under very similar circumstances.

But when he had gained the further bank and stood face to face with his great opponent calmly awaiting him upon the hills, he appears to have become mentally paralyzed. Perhaps then for the first time he became conscious that he had committed an irreparable blunder, that he was playing his adversary's game, and had placed his army in that very situation where Lee most wished to find it.

In prompt and vigorous action lay his only hope of success, whether to prevent, as he still hoped, the Confederate army from concentrating, or whether to hew himself free from the toils that bound him. But what do we find? Instead of rapid movement, delay and irresolution; instead of the impetuous advance of overwhelming masses, a series of feeble and ill-supported attacks. The last was his chief and crowning error. To a daring general much may be forgiven, but from him who has grappled with his enemy and failed to put forth all his strength, even pity stands aloof. Turn to the story of Orthez, read how at the crisis of the fight every man save one single Portuguese battalion was thrust into the battle, and understand with what stern energy victory is compelled.

Although the errors of the Federal generals have been already commented upon, three still remain which ought not to be overlooked. The first of these relates to Franklin's conduct of the attack upon Lee's right. If he intended Birney to support Meade, and that he did there can be little doubt, it was exceedingly bad policy not to have given the duty to Doubleday, whose division belonged to the same corps as those of Meade and Gibbon. Bodies of troops destined to carry out an offensive movement, ought to be, as far as possible, under the same command, so that the officer who directs the advance of the fighting line, and who can best "feel the pulse of the battle," may have it in his power to confirm success by sending forward supports and reserves, at that favourable moment of which he alone can judge. Otherwise, even in highly trained armies, friction and misunderstanding will inevitably occur. "Troops formed for attack," says Colonel Home, "should be in at least three bodies, and these bodies should, under one direction, work for one object, and be closely linked together." Birney's division appears to have acted independently, and not under the orders of General Reynolds, commanding the First Corps, who ought to have had the control of the whole attacking force. A want of concert between brigades and between divisions, and a neglect of the harmonious co-operation necessary to achieve a tactical end, is generally observable in the earlier battles of the American War, and is probably due to the fact that many of the division leaders, brigadiers, and staff-officers had little practical training and possessed but slight knowledge of the art of war. There was too much inclination to indulge in independent enterprise, and to forget that the leader of any body of men depends for success upon the timely and exact obedience of his subordinates. If at Gettysburg, where, it has been said, the South stood *"within a stone's throw of independence,"* Lee's orders had been promptly and strictly carried out, the battle would probably have ended in a Confederate victory.

Again, the whole available strength of the three Grand Divisions should have been brought over the river before the battle opened. The Fifth Corps, which may be said to have constituted the general reserve until 4 p.m., did not commence the passage until after noon; and at 2.15 p.m., when Meade had been repulsed and a counter-attack was to be apprehended, but one division had crossed. A reserve posted at a distance from the decisive point is, as was Bazaine's at Mars-la-Tour and Gravelotte, almost absolutely useless. It has two duties to fulfil: to confirm success, or to ward off counter-attack. The Fifth Corps, placed as it was, could have done neither. It may be, however, that Burnside, ignorant of the presence of Jackson's Corps, thought, when the battle opened, that the services of these 19,000 men would not be needed.

Lastly, a conspicuous mistake was the inaction of the Federal cavalry. Eight thousand men at least, trained to fight on foot, were retained during the whole of the engagement under the shelter of the river-bluffs, or out of range upon the Stafford Heights. Stuart and his brigadiers would hardly have consented to remain idle spectators of the battle, and it is difficult to understand why some regiments at least of the Federal horsemen were not, like their Confederate rivals, converted into infantry for the time being. Two brigades so constituted would have done good service on Franklin's left, and have rendered it possible to utilize Doubleday's division for offensive purposes.

The loss of the armies during the three days' fighting was as follows:—

Federals (1,284 killed)—
Sumner	5,444
Hooker	3,355
Franklin (not including Birney and Sickles)	3,787
Engineers, etc.	67
	12,653

Confederates (595 killed)—
Jackson	3,415
Longstreet	1,894
Cavalry	13
	5,322

Less than 30,000 Southern troops were actually engaged. Eighty-one per cent. of the Federal casualties were reported to have been caused by musketry, five per cent. by shell, fourteen per cent. by round shot; a much larger proportion than is usually due to artillery fire.

After the Federals had crossed the Rappahannock the Confederates prepared to go into winter quarters. Stuart, however, did not long remain inactive. Reconnoitring expeditions were pushed far within the Union lines; the outposts were harassed, the communications disturbed, and convoys captured.

On Christmas Day, three cavalry brigades, led by Stuart in person, passed the rivers by Ely's and Kelly's Fords, and remained absent ten days. During that time they extended their incursion as far as Dumfries, upon the Potomac, fifteen miles N. of Aquia Creek, and to Fairfax Court House, in the vicinity of Washington, thirty miles N.W. of Fredericksburg. Much booty and many prisoners were the result of this expedition; and the fact that at the time 120,000 Federal troops were cantoned between the Rappahannock and the Union Capital, testifies to the skill and audacity of the Confederate leader.[64]

Late in January, Burnside, eager to retrieve his reputation, resolved to cross the river by Banks's and U.S. Fords, and to turn Lee's left. The severity of the weather and the state of the roads made all operations on a large scale impossible, and after three days of intense suffering to his troops he was compelled to abandon the attempt.[65] On the 26th of the same month he was superseded in his turn and replaced by Hooker.

The demoralization of the Army of the Potomac after Fredericksburg was very great. Desertions were frequent, reaching at one period a total of 200 daily, and a large number of officers resigned their commissions. The desponding tone of the Northern press at this time was remarkable, and it is clear that Lee's victory was an unexpected and crushing blow to the people at large. However, the stout heart of the nation quickly rallied, and inspired by the loyal determination of President Lincoln, the Union States turned once more to their apparently hopeless task.

Burnside, by magnanimously taking on himself the whole responsibility of failure, won the regard of the President and

64. It may be questioned, at the same time, if the success of this and similar raids compensated for the waste of horse-flesh incurred. [Evidently Henderson decided later that such raids were worth the cost and the risks involved, for in 1902 he wrote: "... many of these enterprises were much more than forays or reconnaissances. Large bodies of cavalry, accompanied by horse artillery, and stripped of everything which would impede their mobility, operated for weeks, and even months, as detached forces, with specific strategical missions, and the value of their work cannot be overrated" (*The Science of War*, p. 56).]

65. [This was the well-known "Mud March."]

of his countrymen. The popular voice persisted in attributing the disaster rather to the errors of his lieutenants than to the incapacity of the general, and it was not long before he was reinstated in the command he had held previous to his unfortunate campaign.

Sumner, advanced in years and worn out by service, retired from active duty, and in a few months died. Franklin, condemned as the chief cause of the defeat, was removed from the Army of the Potomac and relegated to a less important sphere of action.

One-and-twenty years have passed since the Confederacy fell, but Virginia still bears the scars of the devastating strife; her people are scattered, her broad plantations waste,

> "And where the happy homesteads stood
> The stars look down on roofless halls."

Standing one summer morning not long ago on the vine-clad slopes near Falmouth, and looking south down the tranquil valley of the Rappahannock, so far removed from the ceaseless din of European warfare, I found it hard to realize that those lonely hills had once reverberated to the thunder of five hundred guns, and that the recesses of the forest, stretching away into the dim blue distance, had rung with the rattle of musketry and the fierce cries of struggling hosts. Yet memorials enough of those dark December days are here. The fields whose furrows ran with the blood of Sumner's gallant infantry are high with corn, but the ridge beyond is still crowned with grass-grown trenches; and the village children play among the crumbling earth-works on the Stafford Heights. The quaint red houses of the little city and the trees of the surrounding orchards are scarred with shell and bullet; and in the deep woods beyond the Hazel [Run] are traces of long-deserted camps, and paths cut by the axes of the Confederate soldiery.

Upon the summit of Marye's Hill lie those who fell before it; and more than 15,000 Federals, the harvest of this and many another bloody field, rest beneath the earth where Walton's batteries were arrayed.

The great Republic, mindful of their good service, has done well by those who died for her, and the beautiful and well-kept cemeteries scattered throughout the North are monuments worthy alike of the soldiers' valour and the generosity of their countrymen. Here, above Fredericksburg, under the shade of trees and flowering shrubs, rank upon rank of low white headstones face the west; in the midst a loftier cenotaph enshrines the dust of the unknown dead, and about the regular lines of narrow mounds the turf is green and trim. So sleep the victors, cherished in death with all tenderness and honour, and surrounded by the tokens of a nation's gratitude.

Amid the homes of their own kindred, in the green God's acre of the quiet town, their foemen lie; but though the mourners are more constant, within those gloomy gates are no signs of the lavish care of a victorious Commonwealth; few and poor are the hands that tend the rebels' graves; the grass grows rank about the mouldering headstones, and above them broods the shadow of defeat.

Far and wide, between the mountains and the sea, stretches the fair Virginia for which Lee and Jackson and their soldiers,

"One equal temper of heroic hearts,"

fought so well and unavailingly; a land rich and lovely, but lonely and desolate, bereft of her sons, and mourning always her unforgotten dead. Yet her brows are bound with glory, the legacy of her lost children; and her spotless name, uplifted by their victories and manhood, is high among the nations. Surely she may rest content, knowing that so long as men turn to the records of history will their deeds live, giving to all time one of the noblest examples of unyielding courage and devotion the world has known. This is no place to question the righteousness of their cause, but let none be deceived by the statements of those from whom unreasoning prejudice still hides the truth. It was not to preserve slavery, not in open rebellion against the Federal Constitution, that the Confederates stood in arms, but in the defence of their rights as citizens of sovereign and independent States bound to the Union by a voluntary compact, which they were free to maintain or cancel as they would. Such was the faith of the Southern people. That it was inexpedient may be admitted; that it was illegitimate has not yet been proved.

In the foregoing pages enough has been told to make manifest the splendid fighting qualities of the American soldiers. They were men of the same stock, possessing the same characteristics as ourselves. Those of the South, bred as the majority were to a hardy country life, and hunters to a man, were perhaps more fitted for military service than the greater number of our own volunteers; but their previous knowledge of warfare and its needs was slight. There is, then, no reason why, equal in patriotism, courage, and intelligence, *if knit together by strict discipline and led by well-trained officers*, our own civilian troops, home and colonial, should not, after some short experience of war, excel even Lee's battalions in mobility and efficiency.[66] To emulate their achievements they can scarcely hope, unless a second Lee arise to guide them: for as surely as the legions of Carthage and the Macedonian phalanx derived their invincibility from the genius and inspiration of Hannibal and Alexander, so the Army of Northern Virginia, resolute and daring as were the individual soldiers, owed its matchless endurance to the master-mind and magnetic influence of Lee; that great chief of whom it has been written that, "in strategy mighty, in battle terrible, in adversity, as in prosperity, a hero indeed, with the simple devotion to duty and the rare purity of the ideal Christian knight, he joined all the kingly qualities of a leader of men."

66. [Lord Hartington, who had visited both Union and Confederate armies briefly in 1862, arrived at substantially the same conclusion. In urging passage of reform legislation for the Volunteer forces, Lord Hartington, then Under-Secretary at the War Office, expressed his hope that from the Civil War the English "might learn very many useful things in connection with the services of Volunteers. The army of the North, which seemed to be imperfect in discipline, and which was wanting in *esprit de corps*, had not been found efficient in aggressive warfare; but the army of the Southern States was composed of men animated by very much the same feeling, and drawn from the same class as our Volunteer force." Lord Hartington "felt persuaded" that the English Volunteers were superior even to the Confederates in "physical appearance and strength; in discipline and equipment." If the proposed Volunteer Act were passed, he claimed that 150,000 such men would be available to drive back an enemy "if the soil of England should be invaded" (*Hansard's Parliamentary Debates*, Ser. 3, CLXX [1863], 1698–99). The Volunteer Bill (Bill 108) was passed as amended June 16, 1863 (*ibid.*, CLXXI [1863], 964).]

HENDERSON AND THE AMERICAN CIVIL WAR III

The successive appointments to Sandhurst and the Staff College afforded Henderson greater opportunity for research and writing. In the years that followed, the amount of work he accomplished "was enormous." In addition to preparing and delivering "most carefully thought-out lectures" at the Staff College, he appeared frequently before military societies throughout the United Kingdom. He contributed numerous articles to the *Times* and the *Edinburgh Review* as well as to the various military journals. He received more offers from publishers than he could possibly accept, the new *Military Magazine* even offering him "a guinea a page for anything I like to write."[1] During the 1890's Henderson also devoted what free time he could salvage to the preparation of his major work, *Stonewall Jackson and the American Civil War.*

Henderson's articles and speeches during this period reveal a growing conviction that the tactical lessons of the Franco-Prussian War had been overrated. He continued to take his students

1. Roberts, "Memoir" (see chap. i, n. 2, above), pp. xxvi–xxix.

on periodic tours of the battlefields of 1870,[2] but he himself became increasingly absorbed in the study of the Civil War. In 1891, the same year that his study of Spicheren was first published, Henderson wrote that

> despite the absolute ignorance of war and its requirements which existed amongst the mass of combatants, despite the lack of experience, the tactics of the American troops, at a very early period, were superior to those of the Prussians in 1866. . . . The success with which from the very first the cavalry was employed on the outpost line puts to shame the inactivity of the Prussian horsemen in Bohemia; and, whilst the tactics of the Prussian artillery . . . were feeble in the extreme, the very contrary was the case in the Secession War. . . . Nor were the larger tactical manoeuvres even of 1870 an improvement on those of the American campaigns. . . . Flank attacks and wide turning movements were as frequent in one case as in the other; and not only were the victors of Sedan anticipated in the method of attack by successive rushes, but the terrible confusion which followed a protracted struggle, and for which Prussian tacticians still despair of discovering a remedy, was speedily rectified by American ingenuity. . . . [The American] tactical formations were far better adapted to preserve cohesion than those of the Prussians.

To those who were inclined to discount the lessons of a war waged before the universal introduction of the breech-loader, Henderson pointed out that the Americans, as compared with the Prussians in 1870, "made more careful preparations for attack, were far more zealous to re-form the ranks after every phase of battle, and, whilst developing a broad front of fire, kept within proper bounds the initiative of their company commanders."[3]

Henderson's study of Civil War tactics did not cease with the completion of *The Campaign of Fredericksburg*. His letters to Major Jed Hotchkiss, a former Confederate staff officer, reveal some of the problems that still interested him. "Remember," he cautioned Hotchkiss, who was writing a book on the Civil War,

> what may seem trivial details to you will be exceedingly interesting to soldiers and also to our large army of enthusiastic volunteers. I am now going to be impudent, and suggest what points we should like to hear about particularly.

2. "I go off to-morrow to Germany with a class of officers to visit the battlefields of 1870, and shall be away a fortnight" (Henderson to Jed Hotchkiss, May 9, 1895, Hotchkiss Papers, Library of Congress).
3. *The Science of War*, pp. 129–30.

1. The characters, demeanours, and appearance of your generals.
2. The character of the troops and of their fighting, and of their discipline.
3. The nature of the entrenchments and breastworks constructed.
4. The way in which the fighting in woods was carried out and the precaution taken to maintain order and direction.
5. The way intelligence of the enemy was obtained, and the country mapped. I would suggest you give an example or two of the . . . maps . . . .
6. The methods of the Confederacy [sic] marksmen—the efficiency of their fire and the manner in which it was controlled by their officers—or otherwise.

The more military your book is the better it will go down over here, as, owing to our number of volunteers and our constant little wars, the people generally understand and enjoy all details connected with the grand art of killing one's fellow man.[4]

Henderson was interested in nearly every phase of the military operations of the Civil War. He had written his first book when, as a company officer, he had been concerned primarily with what he termed "Minor Tactics," a phrase used to describe "the formation and disposition of the three arms for attack and defence." But after he had taught a few years at the Staff College his outlook naturally broadened to include more general problems of military policy, organization, and especially that "higher art" of generalship known as strategy or "Grand Tactics," which Henderson defined as "those stratagems, manoeuvres, and devices by which victories are won, and concern only those officers who may find themselves in independent command."[5]

In the following essays, most of which appeared in a posthumous collection of Henderson's writings entitled *The Science of War*, Henderson continued to stress lessons of special significance to the British army. In his article reviewing *Battles and Leaders of the Civil War*, he devoted considerable attention to the problems of a volunteer army. He aired his views on modern cavalry in the twin lectures entitled "The American Civil War." His account of Gettysburg is really an essay on leadership and staff duties, and the Wilderness campaign is treated as a probable preview of wars to come. On virtually every matter of current speculation or controversy Henderson could point to exemplary lessons from the experiences of the Americans in 1861–65.

4. Henderson to Jed Hotchkiss, October 13, 1895, Hotchkiss Papers.
5. *The Science of War*, p. 168.

One of the main military questions of Henderson's day involved the Auxiliary forces, the Volunteers and Militia. There was general and perhaps justifiable concern about the efficiency and capabilities of these organizations, for, while much attention had been given to their assignment in time of war, there were many in England who doubted that untried soldiers such as these could stand up against a large army from the Continent.

Henderson's writings reflect this concern. Unlike his patron Wolseley, who had been content to state that "the armies of raw levies" ought to be taken into consideration when evaluating the generalship of the Civil War,[6] Henderson pleaded for a better understanding of the special problems of the American armies. He stressed the fact that the Civil War had been fought by elements comparable to the Auxiliary forces. "Their experience," he wrote, ". . . will help us to anticipate the shortcomings likely to occur amongst our own volunteers . . . and may enlighten us as to the measures by which these shortcomings may be most readily corrected."[7] While admitting that the Civil War armies had suffered at first from lack of discipline, Henderson also noted that they had improved steadily as the war progressed, until by 1863 they were "in very many respects . . . superior and more advanced in military knowledge than even the Germans in 1870." He believed that the Volunteers could be trained to the point where they were "fully equal" to the Continental armies; in fact, he anticipated that the next great war would be fought largely by armies composed of just such soldiers. Henderson's remarks to a military audience in 1894 seem almost prophetic:

If I see in the future an English general at the head of an army far larger that that which drained the life-blood of Napoleon's empire in the Peninsula, if I see our colours flying over even a wider area than in the year which preceded Waterloo, you may think that I am over-sanguine; but to my mind the

6. Wolseley, "General Sherman," *United Service Magazine*, III, N.S. (July, 1891), 304.

7. See p. 134, below. Henderson wrote this article in 1891, three years after General Brackenbury's memorandum on the subject of "French Invasion," two years after plans had actually been drawn up for the defense of London against such an invasion, and the same year in which the *Stanhope Memorandum*, defining the role of the Volunteers and Militia in the defense of England, was issued (Colonel John K. Dunlop, *The Development of the British Army* [London, 1938], pp. 12–14).

possibility exists, and with it the probability that the forces which are employed . . . will be constituted, at least in part, as were the armies of the American Civil War.

The experiences of the Americans in raising, equipping, and training their volunteer armies in the 1860's Henderson regarded as "one of the most important lessons" to be learned from the war by British soldiers.[8]

Much could also be learned about the handling of cavalry from the Civil War campaigns. During the nineteenth century most European cavalry was still armed with the lance and saber and fought almost exclusively by "shock tactics." Squadrons were massed into dense formations and hurled at the enemy in an effort to crush his forces by the sheer impact of the mounted charge. In the melee that followed, the sword and saber ("cold steel" was the phrase preferred by advocates of "shock tactics") were effective weapons, far handier than rifles at close range. In the Civil War, however, certain conditions prevailed that precluded the handling of cavalry in such masses. Rarely was the terrain adaptable for the traditional charge, while the increased range and rapidity of fire of modern weapons made shock tactics increasingly difficult. Consequently the Civil War cavalry was armed with rifle and carbine and fought for the most part on foot, relying upon firepower for its chief weapon. Contemporary observers had noticed this development without appreciating the fundamental reasons for the change. They claimed that the country was too rough, that the improvised armies lacked time needed to master "a regular cavalry drill," and that the cavalry was wasted in useless raids. They regarded the Civil War as the exception, not the rule, and the few enlightened officers who predicted that the old tactics were doomed in the breech-loader era were not heeded.[9]

8. See p. 257, below.
9. For the observations of those who criticized the tactics of American cavalry see Lt. Col. James Arthur Lyon Fremantle, *Three Months in the Southern States* (New York, 1864), pp. 158, 250–51, 284–85; Fitzgerald Ross, *A Visit to the Cities and Camps of the Confederate States* (London, 1865), p. 31; Lt. Col. H. C. Fletcher, *History of the American War* (London, 1865), I, 256; R. A. Preston (ed.), "A Letter from a British Military Observer of the American Civil War," *Military Affairs*, XVI (Summer, 1952), 51–52; and, at first, Wolseley, "A Month's Visit to the Confederate Headquarters," *Blackwood's Edinburgh Magazine*, XCIII (January, 1863), 27. The most outspoken exponents of the Civil War cavalry were Major Henry Havelock, *Three Main Military*

The controversy over the weapons and tactics best suited to cavalry was still raging in the 1890's, and Henderson approached this issue with the same objectivity and thoroughness that characterized his examination of the qualities of the volunteer soldier. He neither condemned Civil War cavalry for its new tactics nor accepted blindly the arguments of those who shouted that the old-style cavalry was dead. When he wrote *The Campaign of Fredericksburg* he evidently still believed that, in fighting dismounted, cavalry was not "exercising its proper functions."[10] For several years thereafter he thought that unfavorable terrain had been the main reason for the development of dismounted tactics.[11] Further study, supplemented by the experiences of the Boer War, finally convinced him that the rise of mounted infantry was due rather to the increase in firepower which, as early as 1861, "had already become the predominant factor in battle." Although he appreciated the value of mounted infantry and personally was convinced that cavalry "armed, trained and equipped as the cavalry of the Continent, is as obsolete as the crusaders," Henderson refused to side with those extremists who asserted that regular cavalry had no place in modern war. Rather, he believed that the key to the success of the Civil War cavalry was the fact that it had been able to strike "the true balance between shock and dismounted tactics."[12]

Henderson did not believe, however, that the Civil War brand of cavalry could fight on foot as well as regular infantry or could hold its own with European cavalry in "manoeuvring power" and "cohesion." Like Lord Wolseley, he placed little faith in the "military Jack-of-all-arms." Both men hoped instead that a force of mounted infantry comparable to those who rode under Stuart and Sheridan could be provided in Eng-

---

Questions of the Day (London, 1867); Colonel G. T. Denison, *Modern Cavalry, Its Organization, Armament, and Employment in War* (London, 1868); and Colonel F. Chenevix Trench, *Cavalry in Modern War* (London, 1884). Eventually Wolseley moved closer to this point of view. A brief description of the Confederate and Union cavalry is found in Major General John K. Herr and Edward S. Wallace, *The Story of the U.S. Cavalry 1775–1942* (Boston, 1953), chaps. v and vi.

10. See p. 109, above.
11. See p. 210, below.
12. *The Science of War*, pp. 55, 57, 372.

land by the Volunteer cavalry, the Yeomanry. Such a force would supplement the regular cavalry and was needed because the English terrain afforded "even fewer opportunities for purely cavalry combats than Virginia."[13] However, such hopes for converting the Yeomanry into mounted infantry did not materialize. In 1888, two schools for the instruction of mounted infantry had been established for the training of regular army units, principally infantry, but on the eve of the Boer War the Yeomanry were still trained as cavalry proper and retained the sword as an essential part of their armament. Henderson always regretted that British soldiers, by failing to realize the potentialities of mounted infantry, "had overlooked at least one of the lessons of the American campaigns."[14]

Henderson also discussed another significant development in tactics hastened by the Civil War. The aged Duke of Cambridge, regarded in progressive army circles of Henderson's day as something of a military fossil, had grasped the two most important tactical lessons of the Civil War even before the shooting had stopped. Commenting on a lecture by Chesney on Sherman's campaigns in Georgia, the Duke stated that a modern army should contain "masses of light cavalry. Probably the day of heavy cavalry has somewhat passed by." Equally significant was his prediction that "in all future wars, the spade must form a great element in campaigns."[15] Unfortunately, little attention was paid to the use of intrenchments by most European soldiers. The Franco-Prussian War was a triumph of

13. See p. 223, below. Wolseley, too, was against making the regular cavalry learn the duties of mounted infantry. "I, for one, don't believe in the military Jack-of-all-arms, and I feel the result would be a failure; the man would have the efficiency of neither arm" (Wolseley, "General Forrest," *United Service Magazine*, V, N.S. [April, 1892], 5; "General Sherman," *ibid.*, III [June, 1891], 100).

14. *The Science of War*, p. 108. See also A. H. Godley, "Mounted Infantry Training at Home," *Cavalry Journal*, I (January, 1906), 52–55; Dunlop, *The Development of the British Army*, pp. 52–55. The *arme blanche* was abolished for Yeomanry in 1901, the sword being retained for ceremonial purposes only, and in 1909 it was further decreed that the Yeomanry were to be armed only with rifles. Some of the Yeomanry regiments evaded these orders, and at least one regiment, the Northumberland Yeomanry, arrived in France in 1914 with the lance and sword as part of its equipment (information obtained in correspondence with Captain B. H. Liddell Hart).

15. The Duke of Cambridge's remarks are found at the end of C. C. Chesney, "Sherman's Campaign in Georgia," *Journal of the Royal United Service Institution*, IX (1866), 220.

training and organization, but it had been fought along more or less traditional lines. And while the Turks had made skilful use of field fortifications in the defense of Plevna (1878), these operations were regarded as the exception rather than the rule and did not have much influence upon Continental doctrines. Attack, not defense, was the accepted military maxim. Moreover, there was a prevalent belief that troops sheltered by earthworks—or cavalry fighting on foot—somehow lost their spark. French, German, and, to a lesser extent, English military theorists preached the doctrine of the offensive, a doctrine that deliberately played down the use of field intrenchments as partial compensation for the increased power and rate of fire of modern weapons.

In the light of subsequent events, it would have been better for the British had they heeded Henderson's remarks on intrenchments, for here he was years ahead of his time. Although preferring offensive action, Henderson was realist enough to see that trench warfare favored the defensive. He had already made this point in his work on Fredericksburg, and, believing that "the importance of the spade is often overlooked in peace," he warned repeatedly that intrenchments "as a tactical expedient and precaution, and especially as an essential adjunct to attack, do not receive, at field-days and manoeuvres, the attention they deserve." The Boer War further confirmed his contention that intrenchments "play as great a part in modern campaigns as in those of 1861–65 or 1877–78." Henderson considered mobility the best antidote to intrenchments. By the sudden seizure of key tactical points, outflanking maneuvers, and marches against the enemy's line of retreat, the Americans frequently had counteracted the natural advantages of the defensive. It was this ability to maneuver that had enabled Grant to pry Lee out of successive defensive positions in the Wilderness, and Sherman's campaign for Atlanta offered additional proof that "against troops which can manoeuvre earthworks are useless." Henderson did not formulate any special theory based on these observations nor did he advocate unlimited use of intrenchments. (Jackson is quoted as saying, "Armies are not called out to dig trenches, [or] to throw up breastworks ... but to find the enemy and strike him.") But Henderson did appreciate the significance of this lesson, and he advised officers of all ranks to study Grant's operations in the Wilderness because they provided "a better clue to the fighting of the future than any other which history records."[16] Within twenty years many of Henderson's former pupils at the Staff College—Haig, Allenby, Robertson, and Wilson, to name only a few—would have occasion to ponder the accuracy of this forecast.

There is no need to discuss further Henderson's analysis of the Civil War. Everything he says is enlightening, whether it deals with problems of a purely technical nature such as tactical formations for infantry, the position of artillery in attack, or fire discipline or whether it concerns more general problems of discipline, morale, or the relationship between soldier and statesman in a democracy at war. In *The Campaign of Fredericksburg*, Henderson, still a junior officer, took care to harmonize his military views with accepted doctrine. In the following essays he showed greater independence, enabling us to trace the development of his own ideas.

These articles are not without shortcomings. At least one of them, "Stonewall Jackson's Place in History," was "scribbled off" in haste; others contain passages that are redundant and in some cases even contradictory. Many of the military writers cited will be strangers to the modern reader. Time has robbed some of the military questions of their importance.

Despite these obvious shortcomings, however, Henderson's interpretation of the military operations of the Civil War can still be read with pleasure and profit. Few soldiers possess his gift of expression and the time and facilities for detailed research; few scholars can combined an army background with his extensive grasp of military history. In Henderson's view, the campaigns of 1861–65 represented an important development in the evolution of warfare. In our intense interest in the personalities of the Civil War and minute movements on the battlefield, we have often lost sight of the military significance of the war. We have, so to speak, been unable to see the Wilderness for the trees.

Like most English soldiers, Henderson sympathized with the

16. See p. 255, below; *The Science of War*, pp. 68, 341; *Stonewall Jackson* (London, 1906), I, 170; II, 481.

Confederacy, but he did not lose his perspective about military matters. The following letter to an ex-Confederate illustrates his intrinsic honesty:

> I have done my best to be unbiased. Heartily as I sympathize with the South I cannot shut my eyes to the fact that it was better for the world that the Union should have been preserved, and I think Lincoln one of the greatest of men. This you will have to endure, though you will doubtless put me down as an ignoramus.[17]

17. Henderson to Jed Hotchkiss, July 16, 1897, Hotchkiss Papers.

BATTLES AND LEADERS OF THE CIVIL WAR[1]

IV

The War of Secession was waged on so vast a scale, employed so large a part of the manhood of both North and South America, aroused to such a degree the sympathies of the entire nation, and, in its brilliant achievements, both by land and sea, bears such splendid testimony to the energy and fortitude of their race, that in the minds of the American people it has roused an interest which shows no sign of abating. There are few families that did not contribute to swell the rolls of the gigantic armies

1. [This article was written to review *Battles and Leaders of the Civil War* . . . , ed. Robert Underwood Johnson and Clarence Clough Buel, 4 vols. (New York, 1890). It appeared originally in the *Edinburgh Review* for April, 1891, and subsequently was included in *The Science of War*, chap. viii. Present-day practices of scholarship were less rigidly adhered to in Henderson's time, and he often took liberties in quoting the works of others. He did this evidently for the sake of style, therefore no attempt has been made to correct his alterations except in rare instances when the original meaning has been obscured or distorted. His citations have been changed to conform to modern usage. It is of interest to note that Lord Wolseley, who first "discovered" Henderson, and Colonel Frederick Maurice, whom Henderson eventually succeeded at the Staff College, also wrote comprehensive reviews of *Battles and Leaders*. Wolseley's lengthy review of an earlier edition appeared in seven instalments in the *North American Review* (May–December, 1889); that by Maurice was published in the *Journal of the Royal United Service Institution*, XXXIII (1890–91), 1076–87. Both men regarded the publication as an instructive work and stressed many of the same lessons discussed by Henderson in this chapter.]

which stretched in broad line of battle half across the continent; few homes where the voice of the mourner was not heard; few cities that cannot point with pride to the deeds of those who were born within their boundaries. It is little wonder, then, that this intense national interest should have found many channels of expression. The most valuable of these is the stupendous work published under the authority of the Senate, containing as it does every authentic document connected with even the most trivial incident of the war. This official record, however, is inaccessible to the majority of European readers; and its bulk, as well as the nature of its arrangement, renders it valueless to the general public, military or civilian.

The future historian of the great Transatlantic strife—for, excellent as is the work of the Comte de Paris,[2] the history of the Civil War has yet to be written—will find in the autobiographies of many of the prominent leaders, and in the memoirs of others, compiled, as a rule, by members of their personal staff, material sufficient to enable him to explain the purpose of each strategic movement, and to ascribe victories and disasters to their true causes. In addition to these sources of information, and to the numerous histories of individual regiments, almost every State has its Historical Society, and the records of their proceedings contain papers on every aspect of the conflict, contributed by men who took part in the events of which they write. These publications, however, are naturally of a more or less private nature, and their circulation is limited. It has been left to the enterprise of the 'Century' Company to give to the world the reminiscences thus accumulated, and to present them in the most attractive form. Almost without exception, every single article in the four large volumes edited by Messrs. Johnson and Buel is accompanied by illustrations of the ground over which the actions treated of were fought. These illustrations are of a high order of art; they have been executed with a most exact fidelity to nature; and there exists

2. [Louis Philippe Albert d'Orléans, Comte de Paris, *History of the Civil War in America*, 4 vols. (Philadelphia, 1875–88). The Comte de Paris was a member of McClellan's staff during the Peninsular campaign in 1862. He left the Union army in the summer of 1862 because of strained diplomatic relations between Washington and Paris and wrote his massive *History* during the years the Bourbon family was forced to spend in exile.]

no other method which enables the student to realise so readily the features of the battlefields. Without incessant practice, few can reproduce in their mind's eye the landscape depicted on a map; and in any case, as military surveyors have lately recognised, sketches of nature, however rough, are most valuable adjuncts both to maps and reconnaissance reports. The authors of the various papers are of every rank, from the commander-in-chief to the private of infantry; and, taken as a whole, as a picture of war, or a study in tactical science, these volumes are without an equal.

As moral influences remain longest in the memory, and leave the most vivid impressions on the minds of those who have experience of service in the field, it is the moral aspect of war which is invariably the more prominent in personal narratives of marches and of battle. It is in this respect that the 'Century' papers have a value exceeding that of the official accounts of the wars of 1866 and 1870–71. No one can fail to remark the frankness with which the American soldiers speak of the vicissitudes of their campaigns. The simplicity with which they refer to the demoralisation of this brigade, the misbehaviour of that, to the neglect of precaution, to straggling on the march, and to skulking on the field, is in marked contrast to the euphemistic paragraphs compiled by the historical section of the German staff. The latter are so worded as to maintain the invincibility of the German army. It is doubtless considered as essential to impress on successive generations of conscripts that their predecessors yielded neither to panic nor irresolution, as it is unnecessary to inform those who are still their foes how often victory trembled in the balance; and, therefore, we hear but half the truth. On the other hand, with full confidence in the well-proved courage of his people, and without formidable enemies to fear, no American soldier feels either shame or hesitation in admitting that the weakness of human nature prevailed at times over courage and goodwill.

'We heard all through the war,' says a New York private, 'that the army was eager to be led against the enemy. It must have been so, for truthful correspondents said so, and editors confirmed it; but when you came to hunt for this particular itch it was always the next regiment that had it. The truth is, when bullets are whacking against tree trunks and solid shot are cracking skulls like egg shells, the consuming passion in the heart of the average man is to get out of the way. Between the physical fear of going forward, and the moral fear of turning back, there is a predicament of exceptional awkwardness, from which a hidden hole in the ground would be a wonderfully welcome outlet.'[3]

It is in these admissions that the lessons contained in the 'Century' series are exceedingly valuable. Let a man know the exact worth of the instrument he uses, the extent to which its temper may be trusted, the conditions under which it may be expected to fail him, and he will be better armed than the man who looks upon it as an instrument which is to be relied upon under any circumstances whatever. The worth of the instrument with which war is waged depends chiefly on the moral influences to which it is subjected. Armies are not machines, but living organisms of intense susceptibility. It is the leader who reckons with the human nature of his own troops and of the enemy, rather than with their mere physical attributes, numbers, armament, and the like, who may hope to follow in Napoleon's footsteps. To create physical strength in an army is far more easy than to endow that army with moral superiority. 'Many a man,' says the Spanish proverb, 'can make a guitar; few can make music from it.'

'In the "Century" papers,' writes General Maurice, 'you get a sense of dealing with armies of flesh and blood, and not mere war-game counters, unique in my experience.'[4] It is the absence of this element that makes the German histories such terribly dry reading, and, in one important particular, so deficient in instruction. It is its presence in the volumes before us that not only teaches the reader to appreciate the truth of Napoleon's maxim, but suggests the methods in which it may be applied.

There are many questions of importance on which much light

3. *Battles and Leaders*, II, 662.
4. "Battles and Leaders of the Civil War," *Journal of the Royal United Service Institution*, XXXIII, 1082. [In this article Maurice also paid tribute to Henderson. "To those who think that a little practical test is worth a good deal of abstract discussion, perhaps the most convincing evidence I can offer . . . that the [Civil] War is full of instruction of all kinds for soldiers would be to ask those, who have not already done so, to read the admirable short study of the Fredericksburg Campaign, published anonymously about two years ago by an English Officer. Those who have read it will . . . be convinced that, at least from some parts of the war, valuable lessons for present use may be deduced" (p. 1084).]

has been thrown by the events of the Secession War—for instance, the naval operations, mounted infantry, field entrenchments, and the relations of the Government with the leaders of its armies. To these, however, and to other tempting themes, I shall make no further allusion. My present purpose is to examine the history of the war from one aspect only. The great conflict was fought out by unprofessional soldiers, by a national militia, leavened by a sprinkling of regular officers. The armies of both North and South differed little in constitution from an integral portion of our own army of defence. The soldiers were of our own stock. Their experiences, therefore, will help us to anticipate the shortcomings likely to occur amongst our own volunteers should they be called upon to take the field, and may enlighten us as to the measures by which these shortcomings may be most readily corrected.

The bombardment and surrender of Fort Sumter, which first announced to the world that the Northern and the Southern States of America, in Lincoln's homely but expressive phrase, could 'no longer keep house,' took place in April 1861. The regular forces numbered but 15,500, and the greater part of the troops were far away on the Indian frontier. The men held fast to the Union. The officers took the part of their native States, and, under their supervision, armies of volunteers were immediately mustered by either side. Three months elapsed between the assembly of the troops and their meeting on the field of battle, and by both sides this interval was devoted to the work of drill, discipline, and organisation. Men and officers were, generally speaking, without experience of war; and, with the exception of a small minority, the regular officers were utterly ignorant of soldiering. Some few had imbibed a slight knowledge of drill at the military academies which, on the model of West Point, had been established in several of the Southern States. Many had served in the militia and home guards, but these organisations were seldom mustered, and had no more instruction or discipline than was required to quell a riot or take part in a procession.

In the Union States, more intensely democratic than the Confederate, it by no means followed that the more experienced were placed in command. Commissions were given by the suffrages of the men in the ranks, and officers who owed their position to the favour of their former comrades were generally careful not to lose their popularity by the enforcement of an obnoxious discipline. The hold of the officers on their commands was thus of the slightest in the North, and it was but little stronger in the South. The men resented obedience to those who were superior neither in social standing nor professional knowledge to themselves. Of the regular officers available the Confederates made the best use, immediately assigning them to the command of brigades and to posts on the general staff. Nevertheless, despite the presence of these trained instructors, when the two principal armies met at Bull Run, an insignificant stream in Virginia within thirty miles of Washington, the Union capital, on July 21, they both were weak in discipline; and the event goes far to prove that ninety days of camp life were insufficient to give citizen soldiers more than the outward semblance of a regular army.

As regards the actual fighting qualities of the men, the battle was no discredit to either side. Indiscipline was the cause both of the defeat of the Northerners and of the failure of the Southerners to pursue.

'We had good organisation, good men, but no cohesion, no real discipline, no respect for authority, no real knowledge of war. Both armies were fairly defeated, and whichever had stood fast the other would have run.'[5]

'The Federals left the field about half-past four. Until then they had fought wonderfully well for raw troops. There were no fresh forces on the field to support or encourage them, and the men seemed to be seized simultaneously by the conviction that it was no use to do anything more, and they might as well start home. Cohesion was lost, the organisation[s] being disintegrated, and the men walked quickly [quietly] off. There was no special excitement, except that arising from the frantic efforts of officers to stop men who paid little or no attention to anything that was said.'[6]

'At four o'clock on the 21st there were more than 12,000

5. William Tecumseh Sherman, *Memoirs of General W. T. Sherman* (New York, 1875), I, 181–82.
6. *Battles and Leaders*, I, 191.

volunteers on the battlefield who had entirely lost their regimental organisation. They could no longer be handled as troops, for the officers and men were not together. Men and officers mingled promiscuously; and it is worthy of remark that this disorganisation did not result from [defeat or] fear.'

Nor were their opponents in better plight. It is related that as the Confederate President was riding to the field at about four o'clock on the day of the battle, 'he met [such] a stream of panic-stricken rebel soldiers, and heard such direful tidings from the front, that his companions were thoroughly convinced that the Confederates had lost the day, and implored him to turn back.'[7]

Early in the afternoon the Confederates had been driven back by a skilfully conceived movement against their left flank. The generals arrived upon the scene.

'We heard [found] the commanders resolutely stemming the further flight of the routed forces, but vainly endeavouring to restore order, and our own efforts were as futile.

'Every segment of line we succeeded in forming dissolved while another was being formed; more than 2,000 men were shouting each some suggestion to his neighbour, their voices mingling with the noise of the shells hurtling overhead, and all words of command drowned in the confusion and uproar.'[8]

More noteworthy, perhaps, was the inability of the Federal troops, although they had been exercised for the best part of three months in camp, to perform the very trifling marches necessary to bring them into contact with the enemy in good order and in good time.

'The march preceding the battle demonstrated little else than the general laxity of discipline; for with all my personal efforts I could not prevent the men [from] straggling for water, blackberries, or anything else they fancied.'[9]

'General McDowell was anxious to reach Centreville on the 17th, and so to fight on the 19th instead of the 21st, but the regiments, who had only marched from Vienna (six miles),

7. John G. Nicolay, *The Outbreak of Rebellion* ("Campaigns of the Civil War") (New York, 1881), pp. 195–97.
  8. *Battles and Leaders*, I, 210.
  9. Sherman, *Memoirs*, I, 181.

were so fatigued that they either could not or would not push on six miles further the same evening. Their fatigue was partially caused by delays and dawdling, consequent on the ignorance of the rules of marching on the part of the officers, and by the undisciplined state of the troops; and also by the absence of good marching qualities in Americans, and their inability to carry even the slight weights required in light marching order.'[10]

Had the attack been made on the 19th the Northern army would have been opposed by but half the numbers that were present on the 21st.

The disaster of Bull Run roused the Northern States to a truer appreciation of their difficulties, and the President immediately assembled near Washington an army of more than 140,000 men, increased during the winter to 220,000 with 520 guns. In the seven months which elapsed between the first great battle and the second attempt of the North to crush the main army of the Confederates, this force, thanks to the skill and patience of General McClellan, its new commander, gradually assumed the organisation and aspect of a real army. A beneficial change was instituted in the terms of enlistment; the battalions were asked to volunteer for three years or for the duration of the war; and both officers and men set themselves to work more earnestly than their unfortunate predecessors. At the beginning of April 1862, McClellan, selecting the shortest line of invasion, transferred the greater part of his army by sea to Fort Monroe. Richmond, the seat of the Confederate Government, was the objective of the campaign, and so, on the Yorktown Peninsula, already historically famous for the surrender of Cornwallis in 1781, began that series of operations which culminated in the 'Seven Days' Battles,' the defeat of the Federals by Lee, and the withdrawal of their troops to Washington. Whether this repulse was due to the shortcomings of the leader or to the interference of the Government is a question with which we have no concern. The efficiency of the officers and men is the subject of this enquiry and it is only right to state that in the desperate fighting round Richmond, the troops showed far greater stability and endurance than at Bull

10. H. C. Fletcher, *History of the American War*, I, 129-30.

Run. At the same time they had not yet by any means attained either the consistency or the mobility of professional soldiers. The men had not yet acquired the habit of mechanical obedience, which alone makes an army an effective weapon in the hands of its commander. Where duty became irksome it was neglected. Straggling on the line of march was a conspicuous evil. The details connected with sanitation and the care of equipment were generally overlooked, and the health of the troops and the efficiency of their armament suffered in consequence.

Amongst the critics of the campaign are two experienced European soldiers, the Comte de Paris and Colonel Fletcher. The one served on McClellan's staff during the operations, the other accompanied his army as a spectator. The French prince was prejudiced in favour of the North; the Englishman's sympathies were with the aristocracy of the South; but, divergent as were their predilections, they are at one in pointing out that the bonds of discipline in the Army of the Potomac, as the force commanded by McClellan had come to be called, were weak in the extreme. It is in the pages of these eye-witnesses that evidence as to the condition of the Northern troops can best be found.

One of their most serious shortcomings was that on the field of battle the men were accustomed to conduct themselves in accordance rather with the dictates of their own judgment than with the orders of their superiors. At Cold Harbor, where Lee struck the isolated right wing of the Federals, and compelled McClellan to make his famous change of base from the York to the James River, both sides fought with the greatest courage and persistence, and it was not till after seven hours of battle that 50,000 Confederates drove 35,000 Unionists from their strong position on the left bank of the Chickahominy. General Porter, commanding the Northern troops engaged, had exhausted his reserves some time before his line yielded; but fresh troops had been sent across the river by McClellan, and an orderly retreat might have been easily effected, for the Confederates were in no good trim for further action. As it was—'When the crash came no one could stop the current of fugitives: large numbers of men without order, with arms in their hands, left the ranks and walked to the rear, officers were intermingled with them, in some instances leading their companies away from instead of towards the enemy. There was little or no panic; the men said they were weary, had had enough fighting for the day or were in want of ammunition; some squadrons of cavalry attempted to stop the fugitives, the officers threatening them with their revolvers; but all in vain. . . . The regular infantry regiments preserved their discipline better than the volunteers (as they had done at Bull Run), and many, without yielding to the influence of the now widely spread panic, fell, disdaining to fly. As the stream of fugitives, ambulances, and caissons (the guns themselves were abandoned) arrived on the other side of the Chickahominy, they were halted and formed into some sort of order by a line of sentries and strong patrols which guarded the bridge.'[11]

Now this retreat from Cold Harbor did not resemble the rout of Austerlitz or the débâcle of Woerth. It was not the wild rush of a terror-struck mob seeking safety at any price, as at Vittoria or Waterloo. It was not due to lack of courage or to demoralisation, but to defective discipline. But there is something more demanded from soldiers than the struggle for victory; there is the task of preventing defeat degenerating into irretrievable disaster. It was precisely this task that the Federal volunteers were incapable of executing. Men habituated to discipline, when defeat stares them in the face, throng together, for they have imbibed the instinct that only in unity is there safety. They can trust their comrades and their commanders; they have learnt the necessity of mutual support, and the common danger serves but to bind the ranks the closer. But with troops half-disciplined defeat, for a time at least, has the effect of disintegration; order vanishes, and, however great the courage of the individual soldier, a well-trained enemy, vigorous in pursuit, has such an army at his mercy. It is necessary, therefore, that soldiers should be capable of doing more than sustaining the shock of combat. Every battle cannot be a victory, for war is the playground of Fortune. An army must have stamina sufficient to preserve itself from annihilation: and that stamina is given by discipline alone.

11. *Ibid.*, II, 88–90.

Cold Harbor was but the first of the 'Seven Days' Battles.'[12] Day after day the Northern army, falling back through swamp and forest, battled with Lee's victorious troops. But there was no further disaster. Under the most adverse and dispiriting circumstances, the Federals fairly held their own until they reached the strong position of Malvern Hill. There McClellan turned at bay, and repulsed with heavy slaughter the disjointed attacks of the Confederates. No further fighting took place south of Richmond, and the Army of the Potomac was soon afterwards transferred to the river from which it drew its name. It may fairly be asked how it happened that the Federals, after their defeat at Cold Harbor, found strength to show so bold a front, and to administer such sharp blows during the retreat? An army without the discipline to struggle against defeat is an easy prey to a vigorous foe; but the Confederate pursuit was by no means vigorous. For a whole day Lee was baffled by the change of base. The cavalry, who might have cut the enemy's line of retreat, had been despatched to break up his original line of supplies upon the York River, and did not arrive till their opportunity had passed. Maps of the country and guides were wanting. Unpractised generals and staff officers failed to accomplish the combined movements ordered by the commander-in-chief; and even Stonewall Jackson for once broke his own famous maxim 'never to "let up" in a pursuit.'

Having relieved Richmond, Lee turned on Pope, who with an inferior army lay between the Southern capital and Washington. Pope was outgeneralled and outmarched, and the second battle of Bull Run was as decisive a victory for the South as its predecessor. Then followed the Confederate invasion of Maryland; the capture of Harper's Ferry; the drawn battle of the Antietam, where Lee with 40,000 men held his ground against the Army of the Potomac, although it had been recruited to twice his strength; his leisurely retreat; and in December, to close a year of many battles, the bloody repulse of the same Army of the Potomac at Fredericksburg in Virginia. During this period, on one occasion only, at Malvern Hill, were the Federals decisively victorious in any considerable engagement; the remainder of the great actions which stand out as

12. [Mechanicsville was actually the first of the Seven Days' Battles.]

landmarks in the history of the time, if not Southern triumphs, were in no wise disasters.

Now, if there is one thing more than another apparent to the student of the Civil War, it is that the soldiers on both sides were exceedingly well matched in courage and endurance. It is evident, therefore, that if we would discover the reasons of the superiority of the Army of Northern Virginia over the Army of the Potomac we must look further than the temper and spirit of the regimental officers and men. Northern writers have attempted to account for this superiority in a variety of ways. Even Colonel Fletcher has been induced to lend his support to the statement that the agricultural pursuits, the hunting, the riding, the open-air existence of a majority of the Southerners were better adapted to produce good fighting material than the sedentary occupations of the New Englanders. But, as the Confederate ranks were composed in part of town-bred men, so in the Union armies not only battalions, but brigades and divisions, were recruited from the backwoodsmen of Wisconsin and Ohio, from the farmers of Pennsylvania and the lumberers of Maine. Moreover, in all soldierly qualities, the contingents furnished by the crowded cities of the eastern seaboard never at any period of the conflict suffered by comparison with the Western pioneers. There are those, too, who allege that whilst the *gaudium certaminis* inflamed the passionate nature of the Southerner, the colder temperament of the Northern citizen shunned rather than sought the arbitrament of battle; others, citing Jackson's remark that 'he could beat anything with a herd of cattle behind it,' would have us believe that the certainty of finding ample supplies in the hostile camps nerved the resolution of a half-starved soldiery. I am of opinion, however, that in order to discover the secret of the Confederate successes there is no need either to search for nice distinctions in races closely akin, or to appeal to the fact that Lee and his great lieutenant, Jackson, were a head and shoulders above any Union leaders who had as yet appeared. It was not only the genius of its commanders that won the laurels of the Virginian army. Many of its victories were achieved by sheer hard fighting, they were the work of the soldiers themselves, and that the Confederates were able to wrest success from opponents of equal vigour was due to their superior organisation, more accurate shooting, and above all to their stronger discipline. As to the first, the Federal Government allowed the pernicious principle of the election of the officers by the rank and file to flourish without restraint; and secondly the strength of the army was kept up not by a constant stream of recruits to the seasoned battalions, but by the formation of new regiments. Thus battalions which had served in more than one campaign, and had gained experience and discipline, were soon reduced to the strength of a couple of companies; whilst others lately raised boasted a full complement of rifles, but were without officers, commissioned or non-commissioned, capable of instructing or leading their unpractised men. One State, Wisconsin, created no new regiments, but maintained the strength of those she had originally sent into the field; and so 'we estimate[d] a Wisconsin regiment equal to an ordinary brigade. I believe that five hundred new men added to an old and experienced regiment were more valuable than a thousand men in the form of a new regiment, for the former, by association with good experienced captains, lieutenants, and non-commissioned officers, soon became veterans, whereas the latter were generally unavailable for a year.'[13]

The Southerners, on the other hand, early adopted the conscription; the superior officers were appointed by the Government, and the recruits sent to fill the vacancies in the ranks. The President was so strong in the unanimity of his people as to be free from the necessity of conciliating party supporters of the governors of individual States. Few 'political' regiments existed in the South; men commanded because they were competent to command, and not because they could influence votes.

Secondly, 'a great advantage in favour of the Confederate troops was their skill as marksmen. Accustomed as many of them were from their boyhood to shooting with ball [while hunting] bears, deer, and other game, their certainty of aim was acquired by instinct.'[14]

Lastly, as to discipline, whether we agree or not with Colonel Fletcher that the conditions of life in the South were the more

13. Sherman, *Memoirs*, II, 388.
14. William Watson, *Life in the Confederate Army* . . . (New York, 1888), p. 230.

favourable to military excellence, we cannot reject his conclusion that 'the rich planter possessing many slaves entirely dependent on him in regard to food, clothing, medicine, and discipline, acquired habits of command and organisation highly useful to the officers of an army.' Moreover, the population was as distinctly divided into classes as the subjects of a monarchy. The line of demarcation was strictly drawn and the social precedence of the old colonial families was undisputed. The Confederate States were free from the aggressive independence of the North. Obedience was a quality of which they had previous experience. Throughout their history their people had unreservedly committed their political destinies to the members of their great houses, and they followed them now as loyally in the field. Unfortunately for their cause, neither statesman nor soldier was able to persuade them that, however strongly the presence of trusted leaders may assist discipline, it is devotion to duty alone that makes an army always formidable.

So far as history can tell us, no army, however high the standard of education, has become really efficient until obedience has become an instinct, and the presence in the ranks of men accustomed to think for themselves and to reason before acting, however weighty the authority which bids them act, renders the acquirement of such instinct a long process. When soldiers become once imbued with the habit of obedience, then doubtless the more intelligent will be the more useful; but enthusiasm and intelligence will not stand the stress of battle and the hardships of campaigning, unless their possessors have learnt to subordinate their reason and inclinations to their duty. It is open to those in whose ears the very name of discipline smacks of slavery to assert that a powerful instinct of obedience dwarfs the intellect, turns the man into a machine, and rusts his power of reasoning; and in this there is a shadow of truth, but it is only a shadow. If a soldier is never permitted to use his intelligence, never placed in a position of responsibility, allowed neither to act nor move except at the word of command, sooner or later he loses all power of initiative, and there are many occasions in the field where a man must be left to his own unaided judgment. But if the soldier's training is what it should be, his education for individual action will go hand in hand with his acquirement of the habit of self-effacement. It may be difficult to combine two such opposite characteristics, but it is not impossible. The officers of any regular army have the same instincts of obedience as their men, and yet their power of initiative, developed by responsibility, is seldom impaired; and again, the skirmishers of the Light Division, when they had learnt, on the outpost line of Wellington's army, to use their intelligence, and to act without a corporal at their elbow, proved themselves as skilful and as enterprising as the famous voltigeurs of France, and this without losing their capacity for moving like a wall under heavy fire. It is important to be clear on this point, for it is unfortunately to be apprehended that few, except professional soldiers, understand the nature or the value of discipline. They were certainly not understood in America before the war. The sovereign people of the Northern States could create mighty armies, could equip those armies as none had ever been equipped before; but it could not create the discipline of habit —that was deemed unworthy of free men—and in its place relied on the discipline of reason and of patriotism.

From the pages of the Comte de Paris we may learn whether the American product was an efficient substitute for the mechanical subordination of regular troops. Speaking of the sluggishness with which operations were carried on in McClellan's Peninsular campaign, he writes as follows:

'This sluggishness is in a measure enforced on the generals by the nature of their troops. Those troops are brave, but the bonds of subordination are weak in the extreme. It follows, then, that there is no certainty that what has been commanded will be exactly executed. The will of the individual, capricious as popular majorities, plays far too large a part. The leader is obliged to turn round to see if he is being followed; he has not the assurance that his subordinates are bound to him by ties of discipline and of duty. Hence come hesitation and conditions unfavourable to daring enterprise.'[15]

15. *Guerre d'Amérique: Campagne du Potomac, mars-juillet, 1862* (Paris, 1863), pp. 144–45. [Henderson is in error in attributing this work to the Comte de Paris. The real author was the uncle of the Comte de Paris, François Ferdinand d'Orléans, Prince de Joinville. A former rear admiral in the French navy and a well-known writer on naval affairs, the Prince de Joinville accompanied the Comte de Paris and another nephew during the Peninsular campaign. His account of the campaign, bearing the

Again: '. . . Open to impressions, as are all crowds, the men, accustomed to a complete independence of action, were brought to battle actuated by obedience more reasonable than passive, by a sentiment of duty to the State rather than by the instinct of the disciplined soldier, who forgets his own inclination and draws inspiration from that of his officers alone. So, despite their courage, time was necessary to teach them that on ground where the lines of battle were brought close together, it was almost always less dangerous to charge the enemy than to remain exposed to a decimating fire. In default of the mechanism which, in armies well organised, communicates to every man controlling influences as rapidly as do the nerves in the human frame, there were constant failures to transform a first advantage into a decisive success. When certain death awaited the foremost, then it was easy to march slower than the rest—personal courage being by no means equal—it sufficed that only one should hesitate, or be permitted to hesitate with impunity, for that hesitation to become contagious; and so the brave soldier lost his *élan*, the most resolute officer his daring. . . .'[16]

I have already said that an ill-disciplined army lacks mobility. Marching, strange as it may appear to those who have never served with troops in the field or in protracted peace exercises, makes the greatest demands on the subordination of the men and the exertions of the officers. It is no light task to bring a battalion of a thousand bayonets intact on to the field of battle at the proper time. Something more than enthusiasm is required to enable a mass of men to overcome the difficulties of bad weather and bad roads, or the sufferings of fatigue and hunger.

That the American troops, when they entered on the Peninsular campaign, had improved in this respect on the holiday soldiers of Bull Run there is no reason to doubt; but it seems that the marching power of neither army was considerable.

---

signature of "A. Trognon," first appeared in the October 15, 1862, issue of the *Revue des deux mondes*. Almost immediately, it was translated and published in the United States (cf. The Prince de Joinville, *The Army of the Potomac: Its Organization, Its Commander, and Its Campaign* [New York, 1862]).]

16. The Comte de Paris, *Histoire de la Guerre Civile en Amérique* (Paris, 1874), I, 343–44.

The slow progress often made during important operations may be in part attributed to the inexperience of the staff, and in part to Napoleon's 'fourth element,' mud; but we are, nevertheless, justified in believing that it was mainly due to the absence of order and regularity on the line of march. Writing of McClellan's advance, Colonel Fletcher states that 'the whole extent of the road for twelve miles from the scene of action to the lines round Yorktown was encumbered and blocked up by the advancing brigades. Artillery, cavalry, infantry, and baggage were intermingled in apparently inextricable confusion. The rain fell in torrents, the roads were deep in mud, and the men straggled, fell out, and halted without orders, so that the column of route of the Federals resembled much more the line of retreat of a defeated than the advance of a successful army.[17]

In the papers, not the least entertaining and graphic of the series, contributed to the 'Century' by a gentleman who served as a private in McClellan's army, we find the following:

'It was a bright day in April—a perfect Virginia day; the grass was green beneath our feet, the buds of the trees were just unrolling into leaves under the warming sun of spring, and in the woods the birds were singing. The march was at first orderly, but under the unaccustomed burden of heavy equipments and knapsacks, and the warmth of the weather, the men straggled along the roads, mingling with the baggage waggons, ambulances and pontoon trains in seeming [sic] confusion. . . . After leaving Big Bethel we began to feel the weight of our knapsacks. Castaway overcoats, blankets, parade coats, and shoes were scattered along the route in reckless profusion.'[18]

I have stated that the Southerners of the earlier years of the war proved themselves better soldiers than those who served the Union. Both sides showed themselves stubborn on the defensive, but nowhere did the Federals display the dash and energy which characterised the assaults of the Confederates during the 'Seven Days' Battles.' Nor was the superiority of the Southerners less marked upon the line of march. Lee's victories were due as much to sturdy limbs as to stout hearts. But

17. Fletcher, *History of the American War*, I, 439.
18. *Battles and Leaders*, II, 189–91.

the discipline of his troops was insufficient to prevent straggling. It has been recorded that nearly 20,000 men were absent from his ranks at the Antietam. A long series of hard marches and fiercely contested battles, deficiencies of supplies, the want of boots, and the indomitable spirit which induced many wounded and foot-sore men to report themselves as fit for duty when they were incapable of doing a long day's work, had, it is true, a share in creating the great gaps which existed in the muster rolls on the morning of the battle. But Lee's official reports leave no doubt whatever that indiscipline was the real cause of the undue weakness of the army. On September 7, ten days before the Antietam, he reported as follows to the President:

'One of the greatest evils, from which many minor ones proceed, is the habit of straggling from the ranks. It has become a habit difficult to correct. With some—the sick and feeble—it results from necessity, but with the greater number from design. The latter do not wish to be with their regiments, nor to share in their hardships and glories. They are the cowards of the army, desert their comrades in times of danger, and fill the homes of the charitable and hospitable on the march.'[19]

That this vice was by no means unknown even in Jackson's command, which accomplished such remarkable feats of marching as to earn for itself the name of 'foot cavalry,' we find convincing testimony. General Taylor, an old regular officer, was promoted early in the war to the command of a brigade, and was ordered to join Jackson on the Shenandoah.

'The end,' he writes of one of his first marches, 'drew heavily on the marching capacity, or rather incapacity, of the men. Straggling was then, and continued to be throughout, the vice of Southern armies . . . When brought into the field the men were as ignorant of the art of marching as babes, and required for their instruction the same patient, unwearied attention. On this and subsequent marches frequent halts were made, to enable stragglers to close up . . . The men appreciated care and attention, following advice as to the fitting of their shoes, cold bathing of feet, and healing of abrasions, and soon held it a disgrace to fall out of the ranks.'[20]

19. Armistead Lindsay Long, *Memoirs of Robert E. Lee* (New York, 1886), p. 522.
20. Richard Taylor, *Destruction and Reconstruction* (New York, 1879), pp. 36–37.

Within a month his brigade had acquired discipline and cohesion. When he first reported his arrival to Jackson the latter enquired the road and the distance marched that day. 'Keazletown road,' was the reply, 'six and twenty miles.' 'You seem to have no stragglers.' 'Never allow straggling.' 'You must teach my people; they straggle badly.'[21]

It is scarcely necessary to refer for confirmation of these statements to General Hazen's 'The School and the Army in Germany';[22] but it is worthy of remark that this officer, who served with much distinction under Grant and Sherman and also accompanied Moltke to Versailles, whenever he discusses the relative merits of the Federal and the Prussian soldiery, never hesitates to acknowledge that the average mobility of the latter was by far the greater. That he is compelled to draw a comparison unfavourable to the American troops he attributes rather to the ignorance and indolence of their officers than to the indiscipline of the men; but it must not be forgotten that, at the outset of the war, inexperience and physical incapacity were equally destructive of cohesion. To take, for instance, the operations preceding Bull Run: The rank and file of McDowell's army were not all city-bred; many of the battalions were recruited from the lumbering and agricultural districts; many were in great part composed of men in good position and active habits; but want of practice in the mere mechanical action demanded by the orderly progression of a large body of troops neutralised their powers. Now, want of mobility, under any circumstances whatever, is a fatal fault.

In a country like our own, whose limits are small and where railways are as numerous and as closely connected as the threads of a spider's web, it might seem that no more is required than to bring the men up by train and to set them down behind lines of earthworks. But this is an idea which every practical soldier will scout as chimerical. The transit of great masses of troops by rail is, for short distances, less speedy than movement by road, even when everything has been prepared beforehand; the very existence of earthworks will cause the ene-

21. *Ibid.*, p. 50.
22. [Major General W. B. Hazen, *The School and Army in France and Germany* (New York, 1872).]

my to avoid them, to mask his intentions, and to concentrate his troops at some unexpected point. To meet him at that point the defenders must be capable of rapid and orderly movement. Troops that cannot march are but untrustworthy auxiliaries. They cannot be readily transferred to the threatened point. They cannot be relied upon to execute the counter-stroke, the soul of the defence, involving both expedition and endurance. It is useless to call upon them to pursue. And yet, in the face of this fact, marching has been suffered to become a lost art in England; and it is beyond question that, although the picked contingents of volunteers which take part in the Easter manoeuvres excite admiration by the precision of their movements, and by the ease with which they accomplish long distances in trying weather, there are many men in every regiment who, although manifestly unfit for the fatigues of service, are allowed, for want of a physical test, to take their places in the ranks, and are, therefore, absolute encumbrances to mobility. And these men, be it noted, in case of war would not have had the benefit of eight or ten weeks of camp life, as had the men who failed McDowell at Bull Run. How much the Germans in 1870 owed to their constant practice in marching, to their rigid rejection of weakly men, and their sound system of physical training, may be realised from the following instances: Within three weeks of mobilisation, 'the troops had already evinced great marching powers; thus the 5th Infantry Division, under a glaring sun and over unfavourable ground, had made marches of over fourteen miles on four consecutive days.'[23] On August 2, part of the 14th Infantry Division traversed twenty-seven miles. The 33rd Regiment, about the same period, completed in three days a march of sixty-nine miles over mountainous country. At the battle of Spicheren the advanced guard of the 13th Division, when it came into action against the left rear of the French, was twenty-five miles distant from its morning bivouac; and a battalion of the 53rd Regiment took but thirteen hours to cover the $27\frac{1}{2}$ miles that separated it from the field. And be it remembered that in every one of these cases

23. *Franco-German War, 1870–1*, I, Part 1, 111. [The reference undoubtedly is to the official publication of the Historical Section of the German General Staff. The English edition was published in London, 1874–84.]

more than half of the men, drawn from the reserve, had only just rejoined the ranks.

A little later, after the battle of Gravelotte, but still only a month distant from the date of mobilisation, the six army corps which composed the armies of the Crown Princes of Prussia and of Saxony marched for nine days consecutively in their pursuit of MacMahon, in many instances traversing four-and-twenty miles a day. Stonewall Jackson's division, both in the Shenandoah Valley and in the campaign against Pope, often covered an even greater distance in a single day; but no large army in the first three years of the American War went near rivalling this continuous movement of 220,000 men, encumbered with a huge supply train—for the district was barren—and an enormous mass of artillery. That this gigantic effort stripped the Crown Prince of Saxony of one-third of his infantry we know on the authority of Prince Kraft von Hohenlohe.[24] But the missing men were to be found in ambulance and hospital. Stragglers, in the worst sense of the word, there were none. No abandoned knapsacks marked the route; and the absence of all irregularity on the line of march is constantly remarked by those who witnessed the campaigns in France. Every man who was physically fit answered to his name at the evening bivouac. Every man who could carry his rifle was found in his place when the battle opened. Had an American army of '61 or '62 been opposed by one of the same strength disciplined on the German pattern, a few rough marches would have produced an inequality in numbers greatly in favour of the latter.

In the war of 1870–71, the outpost service of the German armies was carried to a perfection which is, perhaps, without parallel in history. In exceedingly few instances were even the smallest detachments surprised; and during the tedious investments of Metz and Paris, ample notice was received of every threatening movement. The standard of discipline and efficiency attained by the German army is that which every European army is now striving to reach, and it is by that standard that the volunteers of America must be judged. I have already shown that they fell far short of German perfection in the mat-

24. Prince Kraft zu Hohenlohe-Ingelfingen, *Strategische Briefe* (Berlin, 1887), II, 230.

ter of marching; and I may now be permitted to add that their enthusiasm and patriotism were by no means proof against the exacting duties of the outposts. Surprises were frequent throughout the war. More than one of the great battles was ushered in by a sudden rush on troops asleep in their tents or in the act of cooking. Many were the instances where the enemy was able to mass almost within rifle shot of the sentries without exciting suspicion of his presence. Little less numerous are the occasions when, of two armies in close proximity, the one withdrew during the night without the other having the slightest knowledge that such a movement was in progress.

It is true that the dense forests which covered the theatre of war were favourable to every kind of secret operation. But the war of 1870 was waged in part in thickly-wooded districts, and there we find not only that the Germans were secure from attack, but, no matter how great the exhaustion of the troops or the danger of the undertaking, that information of the enemy's movements and dispositions was always forthcoming. Every subaltern in charge of a piquet knew his duty. After a forced march or a hard day's fighting no relaxation was allowed. Before the fires of the bivouac were lighted, scouts were moving far to the front. Through the night watches every road and path was traversed at short intervals by patrols; and the earliest light saw stronger parties pushing forward towards the enemy's lines. Had the officers been always as diligent, had the men been sufficiently disciplined to face the fatigues of this arduous service, the American armies would also have been free from the reproach of negligence and surprise.

It is not sufficient for the security of an army that the majority do their duty, as doubtless did the majority of both Federals and Confederates. The carelessness of a few may give the enemy his opportunity. It was the absolute uniformity with which duty was done in the German army that made it so formidable an adversary and so excellent a model.

As to the discipline of the American troops in camp and quarters, in some respects it was decidedly good. Drunkenness was almost unknown, for the men acquiesced without complaint in the orders which forbade the introduction of intoxicating liquors within their lines.[25] Nor was insubordination in the active sense a prevalent crime. But of passive disobedience there was much. The men, in the early days more especially, were accustomed to yield only such obedience as they considered necessary. The officers dared demand no more, and an appeal to the intelligence of the battalions was a far more effective means of rousing them to action than a mere command. At the same time, leaders conspicuous for skill and valour soon won the confidence of the troops, and then their task became an easier one. The soldiers followed the men they trusted without hesitation, and endured the privations he imposed without a murmur. So far their good sense served them; but it did not teach them that instant obedience to orders, no matter by whom they are given or how injudicious they may seem, is more valuable than the obedience which is merely a tribute to superior ability.

'No man but the commander can judge of what is important and what is not.... Soldiers must therefore obey in all things. They may, and do, laugh at foolish orders, but they nevertheless obey, not because they are blindly obedient, but because they know that to disobey is to break the backbone of their profession.'[26]

It is thus that individual intelligence is best exercised; in realising and maintaining the important truth that prompt and entire obedience, mechanical if you will, but none the less powerful, is the mainspring of success.

That the intelligence and patriotism of the American soldiers were not sufficient to keep them in the ranks upon the line of march I have said enough to prove; but in yet another respect these qualities, unbacked by discipline, were found wanting. In the supreme moment, in the hour of battle, when it required no greater acumen than is possessed by the most ignorant of ploughboys to comprehend that every rifle was needed at the front, numbers, that in some cases exceeded those of a strong

25. [This statement cannot be reconciled with the facts. Drinking was prevalent in both armies and greatly increased the problem of discipline. See Bell Irvin Wiley, *The Life of Johnny Reb* (Indianapolis, 1943), pp. 40–43, and *The Life of Billy Yank* (Indianapolis, 1952), pp. 252–54.]

26. Sir Charles Napier, *Remarks on Military Law and the Punishment of Flogging* (London, 1837), p. 13.

division, were found hastening to the rear. At Seven Pines, McClellan states that when Hooker brought up his division about dark he had been delayed 'by the throng of fugitives, through whom the colonel of the leading regiment had to force his way with the bayonet.'[27] At the Antietam, three months later, two Federal army corps, roughly handled in their attack on Lee's left, almost entirely dissolved; and it was reported on the following day that the reduction in one of them was not due only to the casualties of battle, but that a considerable number had withdrawn from the ranks, 'some having dropped out on the march, many dispersing and leaving during the battle.'[28]

Again, at Shiloh, in the spring of 1862, General Buell, coming up to reinforce Grant, who had been surprised and driven back after a desperate resistance, found a crowd of soldiers, which he estimated at near 15,000 men, about one-third of the whole force, cowering under shelter of the river bluffs. And a careful perusal of the numerous narratives of survivors of the battle reveals that unwillingness to remain under fire was no less conspicuous amongst the Confederates.

However sound the discipline, however efficient the police, there are men in every army whom no earthly consideration—neither habit, nor honour, nor fear of punishment or disgrace—will induce to face death and danger on a hardly-contested field. Long before La Haye Sainte had been carried, and while as yet Napoleon's massive columns had been everywhere beaten back, men galloped through the streets of Brussels crying that all was lost. Craufurd's Light Division, making its famous march to Talavera, met 'crowds of runaways; not all Spaniards'[29] significantly adds the great historian. And when on August 18, 1870, the First German army reeled back in confusion from Frossard's impregnable position, it required the presence of the King himself to arrest the flight of the panic-stricken mob in Gravelotte village.

27. George Brinton McClellan, *Report on the Organization and Campaigns of the Army of the Potomac* . . . (New York, 1864), p. 219.
28. *Ibid.*, p. 394.
29. General Sir William Napier, *History of the War in the Peninsula and in the South of France from 1807 to 1814* (London, various editions), II, 178.

At the same time, I cannot recall a single incident from the history of any disciplined army to show that leaving the colours, before the battle was decided, has ever occurred on the same wholesale scale as in many of the great engagements of the American war. Even the insubordinate French regulars of 1870, straggle as they might on the line of march, held staunchly to the eagles in the hour of combat. To find a parallel to the Antietam or to Shiloh we must turn to the operations of Gambetta's levies on the Loire, where whole regiments of cavalry were posted in rear of the line of battle to drive back the fugitives and drive on the laggards.

But there was still another manner in which the vice of insubordination showed itself, a manner characteristic of armies in which the bonds of discipline are frail, and more fruitful of disastrous consequences than the hesitation or misconduct of the soldiery. Insubordination is the most contagious of moral diseases. Let it burst out amongst the lowest, and, if it be not instantly crushed, its poisonous breath will infect the highest. It is no respecter of persons. If the supreme authorities wink at its existence amongst the rank and file, officers even of superior rank will become contaminated. Let men become once accustomed to overlook remissness, and their own respect for discipline relaxes. So it was in France previous to the downfall of the last Napoleon. In 1859 the army had shown symptoms of insubordination. At Solferino the cry had been heard, 'Les épaulettes en avant!'; and when, in July 1870, the Emperor set out on his last campaign, there were those amongst his most trusted subordinates who had lost all sense of duty. Distrust and jealousy reigned in the highest places. *Camaraderie* was a forgotten word; and the absence of concert, the neglect of the most ordinary precautions, and the indifference of the generals to the action or requirements of their colleagues point to indiscipline of the most pernicious kind.

The great fault of the American soldier in the early part of the war was that the obedience he rendered was based on intelligence rather than on habit. He did not resist authority when he considered its demands were reasonable, but when he thought those demands vexatious or unnecessary he remembered his birthright as the citizen of a free State, and refused compliance. This vice spread upwards. As the soldiery followed with reluctance an untried or unpopular leader, as they did not deem it incumbent on them to obey an officer merely because he was their military superior, so the generals, even those next in rank to the commander-in-chief, were not at all times to be relied upon to render cheerful obedience.

'The success of our army [of the Potomac] was undoubtedly greatly lessened by jealousy, distrust, and general want of the *entente cordiale.*'[30]

Even the influence of Lee, trusted and beloved as he was by his veterans, was insufficient to ensure at all times unhesitating compliance with his orders. Jackson, indeed, declared that he would follow him blindfold. But Jackson's conception of duty was not shared by all. Still, the great Virginian captain had rarely to complain of disobedience or lukewarmness. Nor did McClellan, Jackson, or Grant, when once they had established their reputation, find it difficult to exact submission from their subordinates. But far otherwise was it with those in whom their lieutenants had little confidence, who, like Pope and Burnside, were suddenly raised by the caprice of the President above their fellows, or, like Bragg and Halleck, lacked both tact and fortune. To remain loyal to such men was a severe test, and the discipline of many of their officers lost its hold. It is hardly necessary to comment on the extraordinary means adopted by the Federal Government to ascertain the fitness of the military chiefs, the Congressional Committee on the war, before which subordinate generals were examined as to the conduct of their commander, and encouraged to express their opinions on his ability, his strategy and his tactics, with all the freedom that envy could suggest. The 'Century' papers teem with instances of disobedience, of argument, and of hostile criticism on plans of battle; and the reader of such campaigns as that of Fredericksburg, Gettysburg, and Murfreesboro can realise for himself the disastrous results of such breaches of discipline in the higher ranks.

I have written at some length on this question, and for this reason, that, notwithstanding the increased knowledge of war

30. Francis Winthrop Palfrey, *The Antietam and Fredericksburg* ("Campaigns of the Civil War") (New York, 1882), p. 59.

and its requirements, it appears probable that in the future the canker of insubordination is likely to manifest its presence in this form. The spirit of indiscipline is abroad; not only the indiscipline that is bred of self-seeking, envy, or disappointed vanity, but indiscipline conscientiously advocated as the rule of life and morals. 'To render unto Caesar the things that are Caesar's' is a precept, we are told, that has lost its application. There are those who are unpractical enough to believe with Plato that obedience is of value only when based on reason, and to assert that no man need obey a law the enactment of which has not received his individual sanction. However hurtful to the State, such opinions are a hundredfold more dangerous to the army. Without absolute obedience to the spirit as well as to the letter of the law; without a determination on the part of all to render loyal service and cordial support to every authority, however distasteful such a course may be; without the resolution to forego and to check criticism of the acts of superiors, skill and courage are of no avail. A great military writer has recorded in the pages of the 'Edinburgh Review' that, notwithstanding their vast superiority in numbers, wealth, and armament, the twenty millions who upheld the Union were powerless to crush five millions of Secessionists until they had introduced into their armies a sterner discipline. Intelligence and enthusiasm had their trial. For three long years the infatuation of the Northern people in favour of individual freedom lasted, and during those three years the national cause made little progress. At length the scales dropped from the eyes of the Government and the troops. A leader was chosen who throughout his military career had been constant in obedience, chary of criticism, and patient under misconception; but unsparing of condemnation when it was deserved, and impatient of insubordination in his lieutenants.

Under Grant, backed by the unreserved support of Lincoln, whose conversion to the new doctrine of unhesitating obedience was whole-hearted, the Army of the Potomac entered on a new phase of existence and of efficiency. On one occasion only—at the second battle of Cold Harbor, where, after having already lost more than 40,000 men in less than three weeks, the Federal troops were ordered to renew an assault on an entrenched position which had already cost more than 10,000 men—did either officers or men venture to dispute the judgment of the general-in-chief.

Relying on the discipline no less than on the courage of his lieutenants and his soldiery, Grant was able to carry out his policy of wearing out his opponent by incessant attack. The Army of the Potomac was employed as if it was a battering-ram, without consciousness and without feeling. It was a machine, perhaps unskilfully used, but challenging admiration by the manner in which it answered every touch of the manipulator. The lesson had taken long to learn, but it was thoroughly mastered. Brigadiers and colonels forbore to obtrude their advice upon the general commanding. Divisional leaders no longer asked audience of the President to expose the errors of their superior. No leader of an army corps criticised adversely the plan of battle in the hearing of his troops, as Hooker had done before Fredericksburg. The necessity of co-operation and ready support had become apparent; and the truth was at last recognised that even indifferent tactics have a better chance of success, where those who carry them out are in accord, than more skilful strokes if cordial acquiescence in their expediency is wanting. Those who had held high rank in the regular army obeyed, without a sign of mortification, men who had been their juniors in the old service, who had retired after a few years, had been again brought in from civil life, and were now promoted above their heads. The commander-in-chief had no longer occasion to complain, with Marshal Junot in Portugal, that what he wanted was inferior officers who would obey him, and not comrades who thought themselves as good as he was. That knowledge had come to all which at first had seemed the possession of the few, that absolute devotion to duty is a more substantial good than brilliant exploits in the field, and a more enduring glory than the applause of press and populace.

As to the discipline of the troops on the field of battle, I have already quoted the Comte de Paris's statement that, on the part of the Federal troops, there was a decided disinclination to decide the combat with the bayonet.[31] Over and over again,

---

31. [Scheibert had also commented on the reluctance of the Union troops in 1861 to use the bayonet (Major Justus Scheibert, *Der Bürgerkrieg* [Berlin, 1874], p. 30). Heros von Borcke, a Prussian soldier of fortune who served with Jeb Stuart, wrote that "accounts of bayonet fights are current after every general engagement . . . but as far

in the pages of the 'Century' volumes, instances can be found of the line of battle approaching within a hundred, and in some cases within even fifty, paces of the enemy, and there stopping short, not, however, preparatory to retreat, but to seeking cover, and maintaining a fire fight more fruitful in casualties to itself than a determined advance.

That the battalions were capable of maintaining their position under such circumstances is in itself a proof of fine courage. The Germans impress on their infantry the maxim that, when such close quarters are reached, 'if you don't go away the enemy will'; but here were soldiers who refused to move, and who could be depended on to hold out to the last extremity. The Confederates, on the other hand, successful in so many offensive battles, were manifestly capable of the supreme effort necessary to cross the narrow intervening space between the lines, to carry out decisive assaults, and to pierce their adversaries' front.

Mutual confidence is the force that drives a charge home; and this quality is the fruit of discipline alone, for in almost every campaign it is the better-disciplined troops who have displayed the greater vigour in assault. In the war of 1870 the *furia Francese* appears to have passed over to the men of Brandenburg and Bavaria, and in place of the impetuous advance of the long lines of bayonets which made the battle of Napoleon like 'the swell and dash of a mighty wave,' were the isolated counter-strokes of a few brave men whose daring but served to accentuate the irresolution of the mass. Very early in the War of Secession, the Federal commanders, recognising their enemy's disposition to bring matters to a speedy issue, made use of earthworks and entrenchments; the Confederates, at a later period, when the desperate assaults on the Fredericksburg heights taught them that the Northern battalions had at length learnt to follow their officers to certain death, gave up their trust in broken ground and sheltering coverts, and adopted the same means of stiffening the defence.[32]

---

as my experience goes . . . bayonet fights rarely if ever occur, and exist only in the imagination" (*Memoirs of the Confederate War for Independence* [reprinted New York, 1938], I, 63–64).]

32. [In his review of *Battles and Leaders*, Colonel Maurice commented that there was, perhaps, "one misfortune in taking Fredericksburg as a representative campaign;

In 1863, the third year of the war, both armies became equally formidable on the defensive, and—we have it on the authority of officers who took part in the campaigns—the confusion of the earlier fields of battle was no longer seen. After a charge or a repulse the troops rallied quickly to their colours; there was little intermixture of units; and constant practice on the drill-ground enabled the battalions to reform after a hot fight in an exceedingly short time, to take up the pursuit without delay, or to oppose a counter-stroke with unbroken front. Fire discipline, on the other hand, did not exist. Occasionally, when protected by unusually strong defences, the leaders were able to induce their men to reserve their fire to close range, but, as a general rule, whether defending or attacking, the men used their rifles at will.

'The officers were never sufficiently masters of their soldiers to prevent them, when bullets were whistling past, from immediately answering the enemy's fire. In the best Confederate regiments, in the midst of a conflict, the ardent and burning inclination of the soldiers was obeyed rather than the commands of the officers.'[33]

That the fire of infantry should be under the same control as that of artillery is now recognised as the most vital principle of battle tactics; and it is instructive to note that the American volunteers were incapable, at any period of the war, of answering the very trifling demands made by the discipline of an age which rated fire of less value than the bayonet. The official reports of Gettysburg are significant. Amongst 24,000 loaded rifles picked up on the field only a quarter were properly loaded;

---

that of all Lee's earlier campaigns it was the one that depended most . . . upon the effective use of defensive works and a defensive position. So used, it rather tends to confirm an impression in regard to the whole of the war," an impression Maurice considered misleading (*Journal of the Royal United Service Institution*, XXXIII, 1084). Justus Scheibert, the official Prussian military observer with Lee, claimed that Gettysburg marked the turning point in the development of intrenchments: after Gettysburg, Lee faced overwhelming numbers and had no choice but to resort to the tactical defensive (*Der Bürgerkrieg*, pp. 44–50). According to Freeman, Lee's use of intrenchments dated from Fredericksburg; after 1863, "the declining strength of the army forced it more and more to the defensive [and] field fortification became a routine" (Douglas Southall Freeman, *R. E. Lee* [New York, 1935], III, 204).]

33. Edward Lee Childe, *The Life and Campaigns of General Lee* (London, 1875), p. 46. [But see p. 76 above.]

12,000 contained two charges each (both sides were armed with muzzle-loaders) and the other quarter from three to ten.

It has been stated by Lord Wolseley, speaking with the authority of one who is an earnest student of Lee's campaign, and who accompanied the Confederate army in the operations succeeding the Antietam, that at any time during the war a single army corps of regular troops would have turned the scale in favour of either side.[34] This assertion, as I understand it, implies a conviction that 30,000 regulars would, by their superior mobility and cohesion, have given the leader who controlled them the power of assembling superior numbers at the decisive point; in fact—and their own commanders were fully conscious that such was the case—that even at a late period of the war the armies lacked the attributes of regular organisations. Now, the military experience of the combatants was large, their goodwill remarkable; the military code existed in full force, and officers of proved capacity had little difficulty in securing prompt obedience. How was it, then, that not until the war was drawing to a close did discipline become firmly established, and mobility and cohesion characteristic of the troops? The answer is not far to seek. Both Lord Wolseley and Colonel Fletcher have alluded to the extraordinary difficulties thrown in the path of the commanders by the inefficiency of the regimental officers and the staff, but I prefer to appeal to evidence more direct.

'The great difficulty, I find,' wrote Lee to the Confederate President in March 1863, 'is in causing orders and regulations to be obeyed. This arises not from a spirit of disobedience, but from ignorance. We have therefore need of a corps of officers to teach others their duty, see to the observance of orders and to the regularity and precision of all movements. This is accomplished in the French service by their staff corps.' Enumerating then the various appointments necessary, he adds, 'If you can fill these positions with proper officers . . . you might hope to have the finest army in the world.'[35]

'When I compare the 41st Ohio,' says General Hazen, 'with other regiments which worried the patience by their snail-like

---

34. Wolseley, "General Lee," *Macmillan's Magazine*, LV (March, 1887), 328.
35. Long, *Memoirs of Robert E. Lee*, p. 619.

and uncertain movements, I am strongly impressed with the immense loss which our country sustained in consequence of the indolence, ignorance, and shiftlessness of its officers.'[36]

One of the first acts of McClellan, on assuming command of the Union forces in 1861, and also of Grant, on his promotion to the same office in 1864, was to weed the commissioned ranks; the first by a system of examination, the second by the unsparing exercise of his powers as commander-in-chief. During the régime of those able administrators several hundred officers were dismissed [from] the service. These facts speak for themselves. There is no need to produce further testimony. At the beginning of the war, in both the Federal and the Confederate armies, well-trained officers, staff and regimental, were largely wanting. There were few who understood the careful preparations necessary for manoeuvre and movement, few who could enforce the discipline or carry out the details essential to their execution. At a later period many had been suffered to fill the frequent vacancies who had, no doubt, a large acquaintance with warfare, acquired in the ranks, but had not received the training necessary for those who aspire to command. As regards the staff, the number of officers in the regular army of the United States, including those who had retired, did not exceed 2,000; of these, many on the Northern side remained with their own regiments; on both sides many were detailed to command the larger units. Of those who remained available for staff duties few had received special training, and it was some time before they became fitted for their onerous positions. At the outset, the sovereign people, deeming a staff but an ornamental appendage, objected to its formation. McDowell was accompanied by only two aides-de-camp at Bull Run; and when the scanty number employed was at length allowed to be recruited from the volunteers, the majority had yet to learn the very rudiments of their business. And so, throughout the earlier campaigns, the generals were compelled to work single-handed. They were without 'the hundred voices,' the 'hundred eyes,' the 'hundred ears' which alone make possible the skilful direction of the movements of large armies. They had no means of knowing that their orders had been executed as they wished, or even

36. Hazen, *The School and Army in France and Germany*, p. 221.

executed at all. They had no assistance in framing the multifarious instructions which the troops required. The thousand details which must be attended to during every hour of a campaign, if not supervised by the general himself, were altogether neglected.

Those familiar with the campaigns of 1866 and 1870 know how deeply the principle of co-operation has penetrated the spirit of German generalship, with what extraordinary effect it was put into practice, and how the lucidity of the orders issued by the various headquarters simplified its application. But both in Lee's and McClellan's armies the means of ensuring concerted action were defective, and lack of combination was consequently the great tactical fault of almost every battle. The commanders were without the slightest practical experience of the movements of great masses of troops, such as is imparted to the officers of Continental armies in the autumn manoeuvres. Their military life had been passed in the scattered forts along the Indian frontier, where, like General Ewell, a Confederate brigadier at Bull Run and an officer of nearly twenty years service, they 'had learned all about commanding fifty United States dragoons and had forgotten everything else.'[37]

When we read the orders issued by the Confederate headquarters for the assault of the formidable position of Malvern Hill, we cease to wonder at the failure to arrest the Federal retreat from the Chickahominy to the James. The staff who considered the following production sufficient to ensure a combined attack in a wooded country must have been utterly incapable of directing the intricate movements devised by Lee to ensnare McClellan:

'Batteries have been established to act upon the enemy's line. If it is broken, as is probable, Armistead, who can witness the effect of the fire, has been ordered to charge with a yell. Do the same.'[38]

Unfortunately the enemy's line was not broken. Armistead's

37. Taylor, *Destruction and Reconstruction*, p. 37.
38. *Battles and Leaders*, II, 392. [Henderson later wrote Hotchkiss regarding this battle: "I think Malvern Hill one of the very worst fought battles . . . I ever read of, and I am perfectly convinced that you Confederates made such a mess of it because —like we Englishmen too often do—you thoroughly despised your enemies. I am also perfectly convinced that Jackson would never have made a frontal attack on a strong position had he been in command" (Henderson to Jed Hotchkiss, February 7, 1897, Hotchkiss Papers, Library of Congress).]

division [brigade] did not charge. But three of his regiments became involved in action, and, so far as I can ascertain, their shouts were construed as the signal. Two divisions attacked at different times. They were unsupported, and lost 5,000 men without shaking the enemy's hold on his position. It may be admitted that co-operation when in contact with the enemy is no easy matter to bring about, especially in a country covered by swamp and forest. There are, however, three means of overcoming the difficulty: the first, constant communication between the units; the second, thorough reconnaissance of the ground over which movements are to be made; the third, clear and well-considered orders. Now in both the Federal and Confederate armies of 1862 these three points, as a general rule, were disregarded. The staff was possibly too small to attend to the first, too inexperienced to carry out the second, and insufficiently trained to produce the third. When time is pressing and quick decision essential, when an infinite variety of details has to be considered and provision made for numerous contingencies, the framing of orders is a task that demands not only a wide acquaintance with war, but constant practice. It constitutes a special branch in the education of the general staff, and should find a prominent place in the training of all officers, for the power of explaining his intentions so that none can fail to comprehend is as necessary to the subaltern in charge of a patrol as to the leader of an army corps.

Several of the most important battles of the Secession War would, in all probability, have assumed a different aspect had not 'misunderstanding of orders'—a phrase with which the reader of the 'Century' papers soon finds himself familiar, and which is in itself a proof of an ill-trained staff—so frequently occurred. Nor can we fail to remark the inability of even the supreme commanders to inform themselves of the situation of affairs at the front or on the wings. This arose from the fact that 'the general staff did not and could not assist the commander as he should have been assisted. . . . There was not a large personal staff of experienced and talented officers, capable of keeping the general fully informed of the operations of his corps.'[39] The battle of Williamsburg, fought in May 1862, began

39. Alexander Stewart Webb, *The Peninsula* ("Campaigns of the Civil War") (New York, 1881), pp. 182–83.

at seven in the morning. Although he had sent aides-de-camp to the front for the express purpose of reporting, it was one o'clock before McClellan was made aware that his troops were in contact with the enemy. At Seven Pines, June 30, 1862, Johnston, the Confederate leader, remained for several hours in ignorance that a division had taken the wrong road, and the attack he had ordered had not been made. At Gettysburg, in July 1863, as will be seen later, exactly the same error occurred. With every allowance for the close and wooded nature of the country, such a state of things is as inconceivable in an army possessing a well-trained staff as the fact that, although Jackson's flank movement round Pope, in August 1862, was seen and reported by the Federal signallers, not a single cavalry regiment, nor even a single scout, was sent to ascertain the direction of his march; or that Longstreet's division at Seven Pines, ordered to begin the attack, should have crossed a stream by an improvised bridge in single file, when, in the words of one of his brigadiers: 'if the division commander had given orders for the men to sling their cartridge boxes, haversacks, &c., on their muskets and wade without breaking formation, they could have crossed by fours at least, with water up to their waists, . . . and hours would have been saved.'[40]

Lack of reconnaissance was a fruitful source of indecisive successes and of unnecessary loss. Movements were projected and carried out without previous exploration of the ground or selection of the most effective line of advance. Little care was taken to discover the weak points of the enemy's position. The influence of topography upon tactics was unappreciated and the Confederate divisions attacked exactly where the adversary wished them to attack, instead of being directed by staff officers who, riding with the advanced scouts, had already made themselves acquainted with the ground, to the approaches most favourable to the assailant. We may also notice, that, owing to the simple expedient of placing finger-posts at cross-roads, or leaving an orderly to point out the route, being neglected, on several occasions—amongst others at Cold Harbor, South Mountain, and Gettysburg—the Confederate brigades came into action either at wide intervals from the rest of their divi-

40. *Battles and Leaders*, II, 229.

sion, or when the opportunity had passed, or in some cases, not at all.

During the strategic movements designed to bring an army to such a position and in such formation that it can readily exert its whole strength against the enemy, the duties of the staff are no less important than on the field of action. Few but those who have witnessed or studied the operations of large masses of troops can realise the nice arrangements, the constant supervision, the tact, training, and experience necessary to the successive execution of such movements. For all these operations the intervention of the staff is needed, but chaos and confusion are likely to ensue if the officers composing it are but novices.

In more than one respect the Confederate staff was superior to that of the Union army. The intelligence department was exceedingly well organised. The hunters of the South took kindly to scouting and patrol; and the certainty with which, in the dense Virginian woodlands, the Confederate generals received early warning of their enemy's every movement is proof of the priceless service that may be rendered by bold and enterprising horsemen working in their own country. To train volunteer cavalry to move in mass with the speed, the unity, and the precision essential to effective action in the shock of battle is impossible, but the audacity of the Southern troopers, their adventurous and at the same time useful rides within the enemy's outposts, indicate that such troops can still fill an important rôle, especially in a close country, where individual daring and intelligence, as well as superior horsemanship, have free play.

Again, in the earlier campaigns the Confederates were the better marchers. Jackson, in the movement round Pope's right in August 1862, traversed fifty-six miles in two days; Longstreet was little less expeditious. And although the Southern army was unencumbered by the same superfluity of baggage and supplies as the Federal—the troops depending for subsistence on the fields of Indian corn or apple-orchards through which they passed, and the train consisting of a few ambulances and the ammunition carts—for this rapid advance due credit must be given to the staff. At the same time, as regards combinations for battle, the reconnaissance and mapping of the country over which the army was to move, the supply of guides capable of directing the divisions through the swamps and forests—and this in the midst of a friendly population—the arrangements were deplorably deficient.

General Lee's letter, already quoted, conclusively proves that in 1863, two years after the outbreak of the war, the staff had still much to learn. His suggestions for its improvement were, however, unheeded—they were perhaps impracticable, for staff officers cannot be made in a month or two—and Gettysburg was the result. The greatest conflict of the war was the most prolific of blunders. The story of the second of the three days' battle presents a picture of mismanagement that is almost without parallel. On the second day Longstreet, commanding the Confederate right wing, had been ordered to attack at an early hour. The famous position was as yet but thinly occupied, and Lee hoped to crush his enemy in detail.

'At 9 o'clock the general had been expecting to hear of the opening of the attack on the right, and was by no means satisfied with the delay. . . . About 10 A.M. . . . he received a message that Longstreet was advancing. This appeared to relieve his anxiety, and he proceeded to the point where he expected the arrival of the corps. Here he waited for some time, during which interval he observed that the enemy had occupied the peach orchard which formed a portion of the ground that was to have been occupied by Longstreet. . . . On perceiving this he again expressed his impatience, and renewed his search for Longstreet. It was now about 1 o'clock P.M. After going some distance to the rear he discovered Hood's division (of Longstreet's corps) at a halt, while McLaws's division was yet at some distance on the Fairfield road, having taken a wrong direction. Longstreet was present, and, with General Lee, exerted himself to correct the error, but before the corps could be brought into its designated position it was 4 o'clock. . . . The opportunity which the early morning had presented was lost. The entire Army of the Potomac was before us!'[41]

Moreover, the fighting which ensued showed that the mechanism for securing co-operation was still deficient. 'The whole

41. Long, *Memoirs of Robert E. Lee*, pp. 281–82.

affair,' writes Lee's adjutant-general, 'was disjointed. There was an utter absence of accord in the movements of the several commands.'[42] Now, we are all well aware that the difficulties in the way of a double attack are very great. As at Gettysburg, the failure of one wing or the other to move out at the appointed time may be due to the action of its immediate commander; and there are those who will argue that want of co-operation should be charged to the general rather than to the staff. It is true that in the campaigns of 1866 and 1870, notwithstanding the excellence of the Prussian staff, isolated attacks were by no means unfrequent. But there is absolutely no reason why, if the advance of one column is unavoidably delayed, the circumstance should not be immediately reported to the other; and it is the fact that the isolated attacks at Spicheren, Woerth, and Gravelotte were, in every instance, initiated by generals who had full knowledge of the situation, and assumed the sole responsibility of advancing without support. There was no failure of co-operation, for it was deliberately rejected. In the American battles, on the other hand, the generals who sent their troops forward to what seems wanton destruction did so in expectation of support, and in ignorance that support had become impracticable. This ignorance was due to the want of communication between the different units; and the establishment and maintenance of such communications are the duty of the staff. Whilst the American offensive, therefore, during the first phase of the war, was a series of spasmodic efforts, the German offensive of 1866 and 1870 resembles nothing so much as the resistless sweep of a flowing tide, wave after wave hurrying from beyond the far horizon to break in close succession on the shore; and the singleness of purpose, the untiring energy, which were then displayed were due to the training of Moltke's pupils, the officers of the general staff. Never was Napoleon's golden rule, 'marcher au canon,' more zealously obeyed. Superficial students have indeed pointed out that to construe the words of the great soldier so literally as did the Germans is fraught with danger; but they have failed to discern that when the Germans adopted this principle they took care to provide a means of applying it without risk. They understood Napoleon

42. *Ibid.*, p. 286.

better than their critics. They were well aware that their ancient enemy advocated no blind and reckless rush to the first sound of conflict, but that he held it a matter of course that every general, whether of army corps, division, or brigade, kept himself by means of his staff officers informed of the situation at the front, and was thus able to fix the exact point where his presence was most needed. The staff recognised this linking together of the various units to be among the most important of their duties; it had become a matter of routine at the annual manoeuvres and peace exercises; and if the rashness or the ambition of the subordinate leaders sometimes led to irregularity, still the means of assuring co-operation, so deficient at Gettysburg, were always there; and, save when they were wilfully neglected, never failed to bring about the unity of action so essential to success.

I have often thought that the night marches of both Confederates and Federals through the tangled thickets and over the indifferent roads of the Virginian wilderness in May 1864, as well as the ease with which the troops were handled in the many terrible battles that those marches led to, are remarkable instances of the way in which all obstacles disappear before the skill of an experienced staff. There can be no question that the future historian of the war will find little to criticise as regards the interior control of either army in the later campaigns. But, to show the necessity of the members of the general staff being trained to an average pitch of efficiency, I will refer to the last effort of Lee's heroic army to prolong the struggle. After resisting for nearly nine months, with much inferior forces, every effort of the Union commander to breach the long lines of earthworks which covered Petersburg and Richmond, the Confederates, on April 2, 1865, were compelled to abandon their defences. It was still possible to save the army by a movement past the enemy's front, and Lee was able to gain some hours' start. Grant followed quickly, hoping to intercept him. The Confederates were well-nigh starving, and 'Lee pressed on as rapidly as possible to Amelia Court House, where he [had] ordered supplies to be deposited for the use of his troops on their arrival. This forethought was highly necessary, in consequence of the scanty supply of rations provided at the commencement of the retreat. The hope of finding a supply of food at this point, which had done much to buoy up the spirits of the men, was destined to be cruelly dispelled. Through an unfortunate error or misapprehension of orders the provision train had been taken on to Richmond, without unloading its stores at Amelia Court House. . . . Not a single ration was found to be provided for the hungry troops.'[43] Some one had blundered, and the result was the dispersion of a great part of the army and the subsequent surrender of the rest.

The question of the general staff is one of special importance to States who depend for their defence on an army which is not permanently organised for war. It may be possible to assemble armed men in vast numbers, and, if precise arrangements have previously been made, even to concentrate them at a given rendezvous; but to give them mobility—that is, the capacity for moving in full strength and speedily to any quarter of the theatre of war—to enable each unit to take its part in battle, and to secure the co-operation of the whole, a large contingent of specially trained officers is absolutely necessary. Regimental officers, however, efficient in their own line, however familiar with war, are necessarily ignorant of the duties of the staff. I would draw attention to the fact that, notwithstanding the existence of the regular army as a source of supply, two years of actual service had elapsed before either the Confederate or Federal staff could be classed as trustworthy, and I would remind my readers that the German staff owes its perfection not only to a long course of theoretical education under the best soldiers of the day, but to the practical experience of the movements of great masses of troops, acquired at the annual manoeuvres.

I have already pointed out that national characteristics opposed great obstacles to the acquirement of discipline by the American troops; and I may be told that these characteristics being peculiar to America, the lessons of the war do not apply to our own volunteers. But I have also pointed out, and have produced unanswerable testimony in support, that the indiscipline which was the primary cause of the comparative ineffi-

43. *Ibid.*, p. 412.

ciency of the American armies was mainly due to the shortcomings of the regimental officers.

'The men,' says General Palfrey, 'were such soldiers as their officers made them.'[44] Whilst I am ready, therefore, to admit that on this side of the Atlantic indiscipline would find less genial soil, I cannot blink the fact that here, too, the means of checking its growth is wanting.

I do not wish to imply that, had the American officers been well trained, the troops they commanded would have at once assumed the bearing of veterans. To impart to men unbroken to restraint the instinctive subordination which is the lifeblood of armies is the work of time, however efficient the officers; but, as we have seen, with intelligent men, confidence in the ability of their leaders supplies the place of mechanical discipline with extraordinary effect. And even if it be asserted that the individual intelligence and patriotism of our volunteers are sufficient of themselves to prevent the recurrence of the faults and disorders of the Americans, it is not difficult to show that their officers must needs be thoroughly competent. In the Secession War nothing more than discipline was required to give either belligerent an easy triumph. The leading on both sides being equal, the side which possessed the greater mobility and cohesion would have won by weight of numbers at the decisive point. Now the volunteer officers of England and her colonies have a task five-hundredfold more difficult than had Confederate or Federal. To create and to maintain discipline is not in itself sufficient. Their fellow-citizens demand of them that they should be capable of opposing with hope of success, not unprofessional soldiers, but armies led by officers, both staff and regimental, trained to that perfection of efficiency which Prussia was the first to establish and the first to profit by. By those who understand war in the new aspect given to it by German thoroughness the old idea that a man of ordinary courage, intelligence, and activity needed but the habit of command and an acquaintance with drill to make an excellent officer, has long since been repudiated. To lead men in battle is a profession demanding careful education and thorough train-

44. Palfrey, *The Antietam and Fredericksburg*, p. 185.

ing. That the country at large is very far from realising this truth is evident from the reluctance of Parliament to vote the sum necessary for even the most limited field manoeuvres, although in the opinion of every professional soldier, without exception, these practical exercises are the only means of educating its officers. But if our professional soldiers at home lack the opportunities of learning their work that are afforded to the soldiers of every Continental nation, however poor, the volunteers are in still more evil case. Brigade camps, Easter manoeuvres, and schools of instruction are certainly, so far as they go, valuable means of education; but the five or six days, at most, of practical instruction in the business of a campaign afforded are a very poor substitute for the sixty or eighty days devoted annually in every battalion of the French and German armies to tactical exercises. It may, however, be argued that, by passing a professional examination, volunteer officers prove themselves at least sufficiently well trained to secure the confidence, and thus to establish the discipline, of those they command. Of examinations in military subjects I am no blind admirer; they are by no means fair tests of comparative efficiency.

But I acknowledge that examinations are necessary. If the study which they impose does not always lead a man to think, it at least gives or revives a knowledge of useful details. More than all, the attainment of the required standard proves earnestness, and earnestness goes a long way towards winning the confidence of others. Now, the examinations which volunteer officers are called upon to pass before promotion are of so perfunctory a nature, and the standard to be attained is so very low, that they neither compel reflection nor teach details; and so small is the modicum of study and practice they demand that even the most indolent and indifferent are not deterred from facing the ordeal. The examination in tactics is a severer test, a tax on leisure and on application; but it is noteworthy that by no means a large proportion avail themselves of the opportunity of learning something of the science of fighting, and of earning an increased pecuniary grant for their corps. The truth is—and it is time that it was fairly faced—that the weak point of the volunteer forces is the dearth of well-trained officers. No practical soldier who has experience of our citizen troops, either at home or in the colonies, will be found to deny that these troops suffer from the same deficiency which, in their earlier campaigns, rendered the American armies, brave and intelligent as they were, inferior to the European armies of to-day. Yet I am far from believing that the possible efficiency of the volunteer force has been exhausted. On the contrary, I am firmly convinced that, if a higher standard of military training were exacted, a large proportion of both officers and men would welcome its introduction. It is possible that increased demands would thin the ranks; but, even if their numbers were reduced by a third, with a corresponding increase of efficiency, few thinking soldiers would deplore the loss of those whose lack of leisure, inclination, strength, or energy now acts but as dead weight on the zeal of the remainder. If their discipline and leading be defective, providence seldom sides with the big battalions.

In the preceding pages I have said little of the good qualities of the American soldiers. I am none the less convinced that in some respects they were superior, as every army of volunteers will always be, to the conscript levies of European States; and I am of opinion that only sounder training is required to make our own citizen soldiers fully equal to the troops of any possible invader. This is a bold assertion. But if a strict system of rejection were to eliminate from the ranks all, whether officers or men, whom indolence, indifference, or physical incapacity renders unfit to bear arms, leaving only men of the same stamp as those who now, whether at schools of instruction, brigade camps, Easter manoeuvres and the meetings of tactical societies, seize every opportunity to increase their knowledge, we might endure without anxiety even the absence of a large part of the regular army beyond the seas. The zest with which good volunteer officers undertake their duties is in itself sufficient to ensure the rapid mastery of these duties. With work which is half a pastime, wherein they find relief from the routine of their ordinary avocations, monotony has no place. The very freshness of their obligations is attractive of zeal and industry. Nor are they burdened with the thousand details of interior economy which occupy so largely the time and energy of the professional soldier. They can give almost every hour which they devote to their military duties to preparing themselves for the business of a campaign. They can bestow their whole attention on what is assuredly the most interesting, as it is the most important, part of the profession of arms, the leading of troops on the field of battle. The volunteer force, as at present constituted, is an excellent school of physical training. But this is scarcely the purpose for which it is maintained. Give it capable officers, trained company leaders and an educated staff, raise the standard of efficiency, exact a physical test, and it will become the strong arm of a free people, a safeguard against invasion, and an efficient substitute for conscription.

THE AMERICAN CIVIL WAR, 1861–1865

Part I. The Composition, Organisation, System and Tactics of the Federal and Confederate Armies[1]

The subject I have chosen for this paper is one of very wide extent, not only from the vast size of the theatre of war, the enormous armies engaged, the huge loss of life and expenditure of money, the number of battles and engagements, and the long time that the conflict lasted, but also from the many marked characteristics which distinguish American from European warfare, the novelties in organisation and in tactics, and the many new developments and inventions that for the first time made their appearance.

I shall, therefore, have to confine myself to a very brief and general sketch of the history of the war between the Northern and Southern States of America, that four years' struggle which is called by one side the Great Rebellion, by the other the War of Secession.

1. [This chapter was originally a lecture to the Aldershot Military Society. Part I was delivered on February 9, 1892; Part II, one week later. It was published in *The Science of War*, chap. ix.]

Which of these titles is the true one is still a vexed question, and one that it would be useless to discuss, but it is impossible to grasp the significance of certain circumstances and their bearing on the military operations without understanding the cause of quarrel.

At the end of 1860, and in the spring of 1861, the thirteen Southern States separated themselves from the remaining twenty with whom they had hitherto been joined as the United States of America. In thus seceding they exercised a right which they undoubtedly believed was theirs under the terms of the constitution. It is possible that they may have been wrong in their interpretation of these terms, but a close examination of the text of the constitution justifies, I think, the course they took. At the same time, it may be said that whilst clinging to the letter they ignored or missed the spirit. The framers of that charter most certainly never intended that one or more individual States should be free to leave the Union whenever they thought fit to do so.

However, in breaking away from the North, and in forming the Independent Republic of the Confederate States of America, in opposition to the Federal Union, not only did the people of the South believe that they were within their rights, but they also believed that the Government, in refusing to acknowledge their independence, and in attempting to bring them back to the Union by force of arms, acted without warrant or justification.

Whatever we may think of its wisdom, there is no doubt that the strength of this belief accounts for the length and bitterness of the war, and for the extraordinary resolution and devotion displayed by the whole population of the Confederate States.

The primary cause of war was the existence of slavery in the South. Here, in a cotton and tobacco growing territory, where the climate prevented the white man labouring on the plantations, there were 4,000,000 negro slaves. In the North, where the climate was more temperate, and where the greater part of the community was engaged in manufacture, there were no slaves and but few negroes. The constant tide of immigration provided an abundance of labour.

But slavery was only the indirect cause of the split between the States, and it was not the sole cause. For many years the United States had been divided into two sections, on the one side the slave-holding cotton-raising States of the South; on the other, the great manufacturing cities of the east, and the farming and backwoods territories of the west and north.

Between these two sections, corresponding, roughly speaking, to the two great political parties of the country, Republican and Democrat, had gradually sprung up a spirit of bitter hostility, created by collision on questions of the tariff and finance, and intensified by a wide difference in social life and habits. The South, colonised in old days by the English Cavaliers, possessed in its great planters and landowners an aristocracy, and by this aristocracy it was ruled.

The North, colonised by the Puritans and by Dutch traders, was devoted principally to commerce and manufactures. The two sections had little in common; and we are not surprised to find that for many years before the war broke out they had been drawing further and further apart.

The breach between them was widened by the existence of a party in the North who demanded the abolition of slavery throughout the States. This party was but small in numbers before the war. Indeed, to read the Northern newspapers of the period when it began to put forward its doctrines most vehemently, it would seem that slavery had as many advocates in the North as in the South. But, be this as it may, when, in 1860, Abraham Lincoln, who, rightly or wrongly, was believed by the slave-holders to be but a tool in the hands of the Abolitionists, was elected President of the United States, the Southerners, regarding the institution to which they owed their prosperity as menaced with destruction, determined to exercise the right of secession. The North drew the sword in order to punish them as rebels, and by no means with the purpose of giving freedom to the slaves. In fact, in his inaugural message to Congress, President Lincoln distinctly affirmed that the Federal Government had no right to interfere with the domestic institutions of individual States.

The first State to secede was South Carolina, on December 20, 1860. It was followed at short intervals by the remaining

Southern States; but it is worth while noticing that it was not until April 14, 1861, nearly four months later, that the Federal Government, in the person of the President, declared its intention of restoring the Union.

This long delay is curious. It was due to the generous temper of Lincoln, who seems to have believed that time and discussion would heal all differences, and to the aversion of the whole Northern people from civil war. In fact the temper of the North, when secession was first proclaimed, was anything but warlike. The Abolitionists came in for more abuse than the Secessionists. But this temper changed into uncompromising hostility when the South Carolina Militia bombarded Fort Sumter, in Charleston harbour, and compelled a small garrison of United States troops to surrender. The insult to the national flag appears to have made the Northern people at last realise that the Union was in danger. Lincoln's first act was to call for 75,000 volunteers, each State furnishing a certain number of regiments.

Now the first idea that occurs to us, when we hear of the Southern States being declared rebels, is, why did not the Government employ the army and navy, the national police, to punish the seceders? Unfortunately for the Government both army and navy were on the very smallest scale. There were but 18,000 regular soldiers in the United States, and these were serving on the far western border, protecting the frontier settlements against the Indians. And again, the Southern States, directly they seceded, had called out their Militia and formed corps of volunteers, soon amounting to a considerable force. The North, in default of other troops, had to follow suit; and so the great conflict was fought out by hosts of unprofessional soldiers, of whom, broadly speaking, the superior officers alone belonged to the regular army.

Now the fact of the personnel of the armies being for the most part unprofessional had the effect, in the minds of European soldiers, of causing a certain contempt for the American troops. All acknowledge their courage and endurance; but it was generally considered that the war was conducted on unscientific principles, and had, therefore, few lessons worth the learning. A saying, attributed—wrongly, I believe—to Moltke, that the American battles were no more than conflicts between

armed mobs, well illustrates the attitude of European soldiers.²
But a certain number of officers, English, French, and German,
who had the energy to go over and look at the fighting for themselves, amongst them Lord Wolseley, the Comte de Paris, and
Colonels Fletcher and Fremantle, of the Guards, convinced
themselves, from actual experience, that this attitude was
unjust.

Whatever may have been the faults, due to want of discipline
and training, during the first year of the war, 1862 saw a different state of things: and these competent eye-witnesses found
then that, whilst the constitution of the armies and their methods of making war differed very greatly from those in force on
this side the Atlantic, there were hosts of magnificent fighting
men, with leaders who knew the secret of maintaining discipline amongst their volunteers, and of handling them in the
field with skill and with success. They learned, also, that if the
procedure of European warfare was very often departed from,
it was because the nature of the country and the conditions
under which marches were made and battles fought were utterly unlike anything that obtained in Europe. No European general has yet been called upon to carry on a campaign in a wilderness of primeval forest, covering an area twice as large as the
German Empire, and as thinly populated as Russia. Nor has
any Government been obliged to organise enormous armies for
the invasion of such a territory from a multitude of untrained
and inexperienced civilians, with the help of a handful of regular officers, and to manufacture, to collect, and to issue, the
whole of the matériel needed for their use. Moreover, as the

2. ["... it is not true that Moltke ever referred to those campaigns as 'the struggle between two armed mobs' " (Captain F. N. Maude, *Attack or Defence: Seven Military Essays* [London, 1896], p. 9). Even had he made this comment, it would not necessarily mean that this represented Moltke's final views on the subject. There were many soldiers in Europe who would have agreed with such a statement after the fiasco of the first battle of Bull Run (July, 1861). Sir Edward Hamley, for example, described this "Manasses [*sic*] business" as "the greatest joke in the world. Taken with the swagger beforehand, there is nothing in farce equal to it" (Alexander Innes Shand, *Hamley* [Edinburgh and London, 1895], I, 135). As the war progressed Hamley became less critical in his views, and, when his *Operations of War* appeared in 1866, none other than General W. T. Sherman voiced his appreciation of the fact that so eminent a military critic had considered the American campaigns "worthy of being grouped with those of . . . [the] European world" (*ibid.*, p. 188).]

war came to be more closely studied, it was found that every appliance which ingenuity or science could suggest had been brought into play, and that in very many matters Europe had been anticipated. Breech-loaders, repeating rifles, and ironclads were all of them first employed in America; and balloons,[3] torpedoes,[4] submarine mines, the telegraph, signalling both by flag and lamp, were utilised to a degree hitherto unheard of; while the extraordinary engineering works of several of the campaigns have no parallel in European warfare. I may instance one. In the year 1863, the Northern army in the west found it necessary to repair a line of railway 102 miles in length. An infantry division, 8,000 strong, was detailed for the work. The whole of the tools necessary had to be forged by the men, and no less than 182 bridges had to be rebuilt. The work was done in forty days.

Great sieges were also undertaken; and earthworks and entrenchments assumed an importance far greater than had hitherto been the case, and were applied with an ingenuity of which we have not a single previous example.

Even from a very early stage, the cavalry was far more successfully worked as the eyes and ears of the commanders than by either Prussians or Austrians in 1866; and in their mounted force the Americans developed a new arm, whose achievements are one of the most remarkable features of modern campaigns.

Nor were the Americans—the Federals, at least—behindhand in matters of supply. The transport both by land and sea was most efficient. The commissariat [was] exceedingly well managed, and often plentiful even to luxury. All the resources of civilisation followed the troops into the field. Before some of the greatest battles, when the men were lying down waiting for the signal

3. [See Captain F. Beaumont, "On Balloon Reconnaissances as Practised by the American Army," *Corps of Royal Engineers, Professional Papers*, XII (1863), 94–103. Beaumont accompanied McClellan's army in the Peninsula in 1862 as a military observer. He made several ascensions in Professor Lowe's famous balloon and became convinced that the captive balloon "is capable of being turned to practical account" for military purposes. Beaumont subsequently became one of the pioneers of ballooning in the British army (Whitworth Porter, *History of the Corps of Royal Engineers* [London, 1889], II, 190–91).]

4. [See Captain E. Harding Steward, "Notes on the Employment of Submarine Mines (Commonly Called Torpedoes) in America during the Late Civil War," *Corps of Royal Engineers, Professional Papers*, XV (1866), 1–28.]

to advance, the newsboys went down the ranks crying the latest edition of the daily papers; and in certainly one of the camps of the invading armies were posted notices stating that agents were present to arrange for the embalming of those who fell in action, and for forwarding them to their friends in the very neatest coffins at the very lowest prices.

Even those who regarded the American volunteers as indifferent soldiers had always to allow that their courage was beyond question. A few details will give an idea of the resolution with which they fought.

In the four years of the war there were more than 2,200 engagements, including skirmishes.

Of these 149 were important actions, generally involving a loss of at least 1,000 men.

The loss of life during the whole war has been reckoned at something like 500,000. In the soldiers' cemeteries, scattered through the States, 300,000 Federals are known to be buried.

In each of two of the greatest battles, Gettysburg and the Wilderness, the loss, of both sides together, amounted to 50,000 killed, wounded, and missing. In both of these battles the number of those who met their death in the field was larger than the death-roll of the English army during the whole of the Peninsular War, and including Quatre Bras and Waterloo. In both of them the loss of life was greater than at Gravelotte although the numbers engaged were not half so large.

In the month's fighting in Virginia, in 1864, the Federal army under Grant lost 70,000 men.

I have emphasised these bloodthirsty statistics in order to give some idea of the scale on which the battles were fought; and if the 'butchers' bills' were gigantic, the numbers engaged and the extent of the theatre of war were even more remarkable.

At one period the number of men actually serving amounted to 1,500,000; the number of enlistments during the war, on the Northern side alone, was close upon 3,000,000, and this out of a population of 20,500,000.

As regards population, and consequently physical strength, the South was much inferior. There were but 7,500,000 whites to 4,000,000 slaves. The latter were not employed as soldiers by the Confederates, but their labour was of the greatest value, releasing the white men for service with the armies, providing them with food and equipment, and building fortifications. Nevertheless, the strength of the South always fell short of that of the North, and during the last year of the war amounted to very little more than a fourth.

The North was the invader. Twice was her territory penetrated by the Confederates, but never for more than a few score miles, and no single district was occupied for more than a few weeks.

During the four years of the war, on the other hand, nearly every part of the Confederacy was, at one time or another, trodden by the enemy. The theatre of war, then, spread over the thirteen seceding States, and the area of those States contained nearly 800,000 square miles; in other words, a territory as large as the whole of the Continent and more than half as large as India. To India, an English soldier, Sir Henry Havelock-Allan, who witnessed some part of the operations in America, has likened the face of the country over which the armies moved.[5] There are the same great plains and mighty rivers, navigable almost from their source; the same absence of hills; the same great level spaces between far distant mountain ranges; the same scarcity of roads and railways; the same long journeys, not counted by hours but by days and weeks, between town and town. Like India, it is a country of 'magnificent distances.' In two essential particulars there was a difference. The Southern States, generally speaking, were covered by enormous forests, and the climate was not too hot for military operations at any time of the year. In fact, the mud of the South was an infinitely worse obstacle than its fiercest heat.

One point should always be borne in mind in studying the war. The roads of the South, few in number, were infamous in

5. [It is highly improbable that Havelock visited America during the Civil War. He was appointed Assistant Quartermaster General of the forces in Canada in 1867, but during most of the war years he was busy fighting Maoris in New Zealand (1863–65). No mention is made of such a visit in the sketch written by Robert H. Vetch for the *Dictionary of National Biography* (XXII, 826), nor do his writings suggest that he was in the United States during the Civil War. Havelock was an outspoken proponent of mounted infantry, and, in his essay "Cavalry as Affected by Breech-loading Arms," he cited Sheridan's action at Five Forks (April, 1865) and the ensuing pursuit of Lee as evidence of the effectiveness of this type of cavalry (*Three Main Military Questions of the Day* [London, 1867], *passim*).]

quality. The railroads were rough in the extreme, made of the rudest material; but if they were easily destroyed they were just as readily repaired.

I have already said that the higher commands on either side were filled by regular officers. When the war began there were more than 1,200 individuals in the States who had passed through the Military Academy at West Point; of these one-fifth were Southerners and joined the Confederacy. In order to appreciate the work these officers did in the war, it will be well to turn for a moment to the method in which they were educated and trained.

West Point is, undoubtedly, one of the most remarkable of military institutions, and one of the most satisfactory of military schools.[6]

The word school is almost a misnomer. It is, in fact, a university, where the cadets are under military discipline and command, organised as a battalion, and taught military duties in addition to a very severe course of general education. Four years is the length of their stay. During that time they learn their duties, practically and theoretically, as soldiers of the infantry, cavalry, artillery, and engineers, and the discipline is strict as the instruction is thorough. Up to 1850, military history and minor tactics, and the art of war, were not included in the course, but it speaks well for the good sense of the American officers that the majority of them recognised and remedied this deficiency themselves. A society was formed for the study of Napoleon's campaigns, and the greatest of his campaigns were familiar ground.

The practical training of both officers and men was peculiar.

The army was split up into numerous small detachments along the Indian frontier. The largest garrison consisted only

6. [The majority of European observers and critics of the Civil War held West Point in high esteem. C. C. Chesney had visited the U.S. Military Academy while Lee was superintendent and had formed a favorable impression of that institution (*Campaigns in Virginia, Maryland, ...* [London, 1864–65], I, 50 n.). Lieutenant Colonel Henry Charles Fletcher, an observer with both armies in 1862, submitted a special report in which he urged the creation of a comparable school in Canada (*Report on the Military Academy at West Point, U.S.* [n.p., n.d.], p. 24). For other favorable comment on West Point see F. P. Vigo Roussillon, *Puissance militaire des États-Unis d'Amérique d'après la guerre de la sécession* (Paris, 1866), p. 23; General DeChenal, *The American Army in the War of Secession* (Fort Leavenworth, Kans., 1894), p. 133.]

of a few troops or companies. It was seldom that a colonel had the whole of his command under his hand at one time. Many of the posts, isolated in the Western deserts, were held only by a handful of men. 'During my army service,' said one of the great Confederate generals, 'I learned all about commanding fifty United States dragoons and forgot everything else.'

Now although this system of dissemination and detachments prevented the senior officers and the staff from gaining any practical experience of the movements of troops in large bodies, or learning how to work the three arms in combination, it had a good side as well as a bad one. Not only were the officers in command of the numerous posts compelled to act on their own responsibility, but command had often to be exercised by those of junior rank; and the constant expeditions against the Indians, sometimes employing a thousand men, but more often a troop or company, increased the self-reliance and habits of command already acquired in time of peace. I think there is nothing more remarkable in the history of the war than the capacity for accepting responsibility, and acting on their judgment, shown by the regular officers of every rank. And they were cool-headed enough to draw the line between initiative and rashness. In the very first great battle, that of Bull Run, the quick initiative of two young brigadiers, Jackson and Evans, who had neither of them commanded even a battalion in peace, practically saved the day for the South. The capacity for accepting the responsibilities of independent command is the more remarkable when we learn that by far the greater number of those who rose during the war to the command of army corps and armies had held no higher grade in the old service than that of captain.

Of the other officers holding high rank on both sides, those who went into the war straight from civil life, some did excellent service even in command of army corps and on the staff; but many, who were too rapidly promoted, failed ignominiously, and in many cases the purchase of experience was a very costly business for the cause they served. As a proof of the value of the training given by a military life, I may mention that one of the few foreigners—and there were many engaged—who was promoted to a high command, had once been in the English army, holding the rank of corporal in the 41st Foot. Cleburne's division was by no means the least efficient in the Confederacy, and he himself attributed his rapid rise to the habits he had acquired in the ranks of his old regiment, and prided himself that he at least knew how to keep his white facings cleaner than those of any other general in the Southern army.

Of practical experience the senior officers had nearly all had a good deal in the Mexican War of 1846–47, and the majority of all ranks had seen service against the Indians.

I have said that one-fifth of the West Point graduates resigned their commissions in the regular army, and offered their swords to the Confederacy. The rest of the army, officers and men, held to the Union. It is interesting to note how the services of the officers were utilised by either side.

The Confederate President, Jefferson Davis, was himself a 'West Pointer.'[7] He had served with distinction in Mexico; and afterwards as Secretary of State for War he had had many opportunities of learning the capabilities of the senior officers of the army. This knowledge he turned to good account. His selections for command were judicious in the extreme. Regardless of seniority he chose the man he thought best suited for the billet, and his choice seldom belied his judgment. All those regular officers who joined the Confederacy were placed in high command, or on the General Staff; some took over volunteer battalions, but as a rule at least a brigade was found for them.

In the North, on the contrary, the regular officers were at first somewhat overlooked in favour of the volunteers, and nearly 600 captains and subalterns were retained with their own regiments.

Nor did Mr. Davis, taught by his military experience, desert a commander because he had been unfortunate. He knew too well how much luck has to do with military operations, and so

7. [The fact that a man was a "West Pointer" was quite important to Henderson. In his letters to Hotchkiss, he occasionally inquired whether a particular officer was a graduate of the Military Academy (Henderson to Jed Hotchkiss, April 11, 1895). Occasionally he went too far in stressing the importance of a West Point education: he described General Richard Taylor as a West Point graduate (*Stonewall Jackson*, I, 312), when in fact Taylor had received his education at Harvard and Yale and had no formal military training (Douglas Southall Freeman, *Lee's Lieutenants* [New York, 1944], I, 349). See also p. 147 above.]

long as a commander showed skill and resolution he was maintained in his position. Both his greatest generals, Lee and Stonewall Jackson, met with ill success in their first independent command. Once only did the Southern President depart from this rule, in relieving General Johnston in 1864, and then he committed an irreparable mistake.

Very different was the procedure in the North. Neither Abraham Lincoln nor his Secretary at War had any previous knowledge of military affairs, but, notwithstanding, they not only attempted to dictate to the generals in the field, but settled for themselves who those generals should be. If their efforts to direct military operations were disastrous, as Lord Wolseley has pointed out, their efforts at selection were little better. The voice of the people exercised much influence over their choice, and generals who had won trifling successes over inferior troops were preferred to those who had proved themselves worthy, if unsuccessful, opponents of the best generals of the Confederacy. Commander succeeded commander with startling rapidity. The chief army of the North, that which was engaged in Virginia, was commanded by no less than six different officers, each one of whom, except the last, was degraded for ill success. At the same time, volunteer generals who commanded great political influence were retained in their command, despite the constant exhibition of the most glaring incompetency. Later in the war, the President and his advisers, and even the sovereign people, learned wisdom. In General Grant they found at last a successful leader, and they forbore to interfere with him. He was allowed to choose his own subordinates, and to dismiss those who were incapable as he pleased. The power entrusted to him he carried out with no sparing hand. In almost the last battle, during the night which intervened between its phases, a corps commander, who had served with much distinction throughout the war, but had shown himself somewhat deficient in energy at critical moments, was summarily relieved of his command; and this not by Grant himself, but by Sheridan, the general in immediate charge of the operations.

Not the least interesting study connected with the war is that which concerns itself with the individual commanders on either side. Their personal histories are all well known, and a general survey of them brings a number of interesting facts to light. As a rule they were young. Very few of those who made great names for themselves were more than fifty. Stuart and Sheridan, the two great cavalry leaders, were under thirty when the war broke out, and several of their most distinguished lieutenants had no more than four or five years' service. The most dashing Horse Artilleryman in the Confederacy was twenty-three when he was killed; and one of the best cavalry divisional commanders on the Federal side, General Mackenzie, did not even leave West Point until the war was nearly half over. There are also some interesting facts bearing on the question of training and experience. I have already alluded to General Cleburne. We should scarcely expect to find that some three years' service in the rank of corporal in an English regiment fitted a man to command a division in the field. Some of the volunteer officers, moreover, who joined without any previous military knowledge whatever, made dashing and skilful leaders, notably General Terry on the Northern side, and Forrest on the Southern. The latter, who proved himself a most able tactician, would most certainly have failed in any written examination for promotion. He could read or write only with the greatest difficulty.

Again, several of the most famous generals had, for a long time before the war, severed themselves from all connection with command and with the service.

Longstreet, one of the very ablest officers in the South, came from the pay department. Grant had been regimental quartermaster, had left the army and been employed as a clerk in a tannery. Sherman had only thirteen years' army service, and had since been lawyer, banker, and professor in a military school; D. H. Hill had been professor in a university, and afterwards a lawyer; McClellan, president of a railway company; and Stonewall Jackson, perhaps the greatest soldier of them all, had served but four years in the Artillery, and for the ten years preceding the war had been Professor of Mathematics and Artillery in the Military Institute of Virginia. Another Confederate general was at the same time a bishop; and he was not the only ecclesiastic who, having left the army for the church, resumed his former trade when the war broke out. Lee's chief of artillery, General Pendleton, was an Episcopal clergyman, who, it is said, condoned his relapse by always prefacing the command to fire with the words, 'The Lord have mercy on their souls!'

None of these officers appear to have found the want of that practice and training which are given by immediate contact with the troops. Grant and Sherman both tell us in their memoirs that when they first took command they were ignorant of the drill then in use. But it is possible that had they never severed their connection with the army, their success would have been far more remarkable than it was. Many others in like case with themselves failed ignominiously. They were but the exceptions that prove the rule. Without character and capacity, physical and moral courage, coolness, and self-reliance, it is impossible that a man can become a great soldier. But, however strong he may be in the possession of such qualities, study and practice can never be anything else but beneficial. In some degree they are essential; and those who are not exceptionally gifted should take to heart the opinion of one of the most experienced of the Confederate generals. 'Conscientious study,' he says, 'will not perhaps make them great, but it will make them respectable; and when the responsibility of command comes, they will not disgrace their flag, injure their cause, nor murder their men.'

Now as to the regimental officers and men.

The private soldiers on both sides were drawn from all classes of society. Men of the best breeding and culture in America, of high education and great wealth, marched shoulder to shoulder with small farmers and clerks, with mechanics and labourers. In the North there was a proportion of men who enlisted merely for the sake of high bounties, and a number of foreigners. In the South a proportion were conscripts; but, on the whole, the patriotism and good-will of the armies were undeniable.

The number of foreigners in the Federal armies has been greatly exaggerated, but there were whole divisions of Germans, and on both sides there were battalions and brigades of Irish. It may be interesting to mention that whilst the Irish were everywhere counted as excellent soldiers, the Germans fell short of such a reputation.

The *moral* of the armies, leavened by the presence of men of intelligence and high principle, was necessarily good. Crime was practically unknown; of insubordination there was very little, but, at the same time, the standard of discipline was never a very high one. It appears to have depended altogether on the personal character and capacity of the commanding officers, and even in the best regiments it seems to have been impossible to exact the same strict regard for duty as in a professional army. The truth is that neither officers nor men possessed the *habit* of obedience. They were willing enough, patriotic enough, and as plucky as soldiers ever were, but they could not be depended on to obey under every circumstance, no matter by whom the order was given. Obedience was not an instinct, and good-will did not prove an efficient substitute for the machine-like subordination of the regular. The question of American discipline is a difficult one. I do not know of any writer on the war who discusses it at length, and all direct information on the subject comes from stray remarks and admissions that might easily pass unnoticed. But at the same time it an interesting question, especially to those who may have to deal with our own Volunteers, and perhaps it will not be out of place if I give the impressions that a long study of the history of the war leaves on my own mind. In the first place it seems that the men wanted a deal of humouring, and the regimental officers also. Mistakes had to be overlooked and ignorance excused. Marks of respect to rank and the ordinary etiquette of an army had often to be dispensed with, and it was injudicious to interfere between the regimental officers and their men. Freedom of speech could not be checked, and there was much familiarity between even the generals and the privates. Still, taking into consideration the democratic constitution of the States, it is possible that these things might have existed, and 'the thinking bayonets,' as their leaders were so fond of calling them, have been as reliable soldiers as the best of European troops. But there are certain facts which show, I think, that 'the thinking bayonets,' however high their spirit, would have done better had their habits of obedience been so ingrained as to rise superior to all personal feelings whatever, whether of danger, hunger, or fatigue. These facts are as follows:—

1. The very prevalent habit of straggling from the ranks on the line of march which seems to have existed certainly for the first three years, and to have existed unchecked, and we can understand how much the generals must have been hampered in their operations by their uncertainty as to the number of men they could count on to reach a fixed place at a fixed time.

2. The very indifferent manner in which the infantry outpost duties were carried out, at least for the first two years of the war. Instances of surprises, not of small parties, but of whole armies, were numerous. Of course in the forests of the South outpost duty was most exacting, but that more than one great battle should have been begun by the rush of a long line on troops surprised in the act of cooking, or asleep in their tents, seems a proof that sentries and patrols were not so vigilant as they should have been.

3. The absolute want of control over the fire of the men. The only symptom of fire discipline was that the men could generally be induced to reserve their fire to short range where they were well sheltered and the enemy was advancing without firing. Directly the bullets began to fly the men 'took charge.'

These shortcomings bear out Lord Wolseley's opinion that the presence of a single army corps of regular soldiers would have turned the scale in favour of either side.

Before I turn to the actual campaigns there are two circumstances bearing very strongly on tactical efficiency which should be noticed.

The Southern States were a wilderness of forest, swamp, and river. Game was abundant, and the great hunting grounds were free to all. Sport in all its forms was the regular pastime of the whole population, and the men of the South were accustomed from childhood to the use of the gun and rifle. 'Nine-tenths of our men,' says a Confederate officer, 'were excellent shots and practised judges of distance.' A book written by an Englishman, who served with a Confederate regiment, tells us that on Enfield rifles being issued to the men the first thing they did was to knock off the elevating back-sight. They judged distance by instinct and wanted no mechanical contrivance to assist their aim. Now in the North, the people of the Eastern States and the foreigners who enlisted knew very little about shooting;

and I do not believe that they had much ball practice during their service, except on the field of battle. It was by these troops that the Confederates were opposed in Virginia, and the superior marksmanship of the Southerners had undoubtedly much to do with their long succession of victories. In the western quarter of the theatre of war on the other hand, in the Mississippi Valley, the fighting was of a much more give and take character; in fact here the Northerners were more often successful. To this result their superior numbers had doubtless something to say. But it was probably due rather to the characteristics of the Northern troops engaged. The men were drawn from the Western States; and among them were many farmers or backwoodsmen, as expert with the rifle as their opponents.

The second circumstance is that the Southerners were a nation of horsemen. Fox-hunting flourished in many parts of the States, and no white man ever walked when he could ride. In the North the very contrary was the case. Horsemanship was practically an unknown art, and had it not been that the regular cavalry regiments were available for service with the Federal armies, it is probable that the superiority of the Southern troopers in the first two years of the war would have been more marked than it actually was. It will be seen, then, that the Confederacy, inferior in numbers, in resources, and in wealth as it was, started with two great tactical advantages, advantages which it took the North a very long time to overtake. But at the same time there were counterbalancing advantages on the side of the Federals. Their artillery was always superior to that of the Confederates, both in material and in personnel. The forty-eight batteries of the regular army served as models to the Northern volunteers. One regular battery was grouped with three manned by volunteers, and the latter quickly profited by the example set them. Again, the supply of horses in the North was practically inexhaustible, whilst in the South there was always the greatest difficulty in providing the cavalry and artillery with remounts.

No preliminary sketch of the armies would be complete without a reference to the mounted arm. Wooded and close although the country was, this branch of the service soon showed its value, and at the end of the war the strength of the Federal cavalry was over 80,000. As is well known, the American horse resembled mounted riflemen rather than ordinary cavalry. Although they were quite capable of charging, and were just as efficient on the outpost line as the best of European cavalry, the principal part of their fighting was done on foot. And this was not because they were indifferent riders or were ill-trained —far from it—but because of the close and difficult nature of the country. Lord Wolseley has been rather severely criticised in America because he has stated that on the theatre of war there was no ground suitable for cavalry engagements as we understand them in Europe.[8] His critic asserts that there was a large extent of such ground. I believe, however, that the ideas of Lord Wolseley and his opponent as to what sort of ground is suitable for cavalry work differ very greatly. The former was probably thinking of the great plains of France and Germany, stretching away for mile upon mile without the least obstacle to free movement. The American was probably thinking of the clearings in the Southern woodlands, spaces very circumscribed in comparison with the rolling downs of Mars-la-Tour. There is a set of maps, in minute detail, of the scene of many of the American campaigns, and it is hard to find on any of them any locality so unencumbered with woodland as to afford a satisfactory arena for the ideal cavalry battle. I have carefully measured the scene of the battle of Brandy Station, the greatest cavalry engagement of the war, and I can find no open ground, free from wood or stream, more than a mile square. Besides, the large clearings which did occur had been made by the farmers, whose barns and fences considerably interfered with the manoeuvres of the cavalry. A personal knowledge of Virginia has convinced me that it is a country just as unsuited to ordinary cavalry fighting, as we understand it, as England itself. It is quite possible that had the country been open the Americans would, before the war had ended, have possessed a splendid force of cavalry pure and simple. But, as it was, in such a country, there was little use for such

8. [For criticisms of Lord Wolseley's views see W. T. Sherman, "Grant, Thomas, Lee," *North American Review*, CXLIV (May, 1887), 437–50; General James B. Fry, "Lord Wolseley Answered," *North American Review*, CXLIX (December, 1889), 728–40.]

a force, and the cavalry leaders very quickly discovered that their men were far more valuable as mounted riflemen.

The next interesting question is: to what degree did these mounted riflemen combine their two characteristics? Were they good infantry and at the same time good cavalry?

On the outpost line they were most efficient. The extraordinary raids they made on communications and magazines were a distinctly new feature in war. They stormed earthworks, they captured cities, and they even went so far as to attack and capture gunboats, but, at the same time, when dismounted they were not considered as efficient as the ordinary infantry, and as cavalry I do not believe that they would have been able to cope with good European troops in open country. But they were admirably adapted for all mounted work in the Southern forests, and no European cavalry would have been able to touch them on their own ground. The American idea, to this day, however, is that good mounted riflemen are more than a match, on any ground, for European cavalry.

The chief staff officers on both sides were recruited from the regular forces, but the enormous armies demanded a very large reinforcement from the volunteers. I need hardly say that an army of 18,000 men, scattered all over the western prairies, could scarcely be expected to supply any large number of well-trained staff officers, and, at first, whilst the staff was new to its work, many were the blunders which were due to the inexperience and ignorance of those who composed it. Later in the war things were very different, and in many of the campaigns, such were the celerity and precision with which enormous masses of men were moved, handled, and supplied, that the first thing that strikes us is what a remarkably efficient staff the generals must have had.

As to armament, I may add that the infantry on both sides were armed with muzzle-loading rifles. The guns also were muzzle-loaders, rifled and smooth bore. The Northern cavalry, after the first year, carried breech-loading and repeating carbines, as well as sabres and revolvers. In the arms of the Southern troopers there was little uniformity. Many of the regiments were supplied with carbines, but others carried long rifles. There were regiments of lancers raised by the Federals in 1861, but they were soon converted into ordinary dragoons.

After the bombardment and surrender of Fort Sumter had brought matters to a crisis, and President Lincoln had called out his volunteers, the Federal Government set to work to devise a plan of campaign; and there are certain geographical and political features which must be made clear before that plan can be properly understood.

1. The long seaboard of the Southern States, and the small number of harbours.

2. The Mississippi river, running from north to south right through the States, and dividing Texas, the great cattle-raising State, from the remainder of the Confederacy.

3. The position of the north-western corner of Virginia, running up into the heart of the North, and contracting the isthmus which, south of Lake Erie, joined the eastern and western portions of the Northern territory to a neck little more than 100 miles in width.

4. The position of the rival capitals, Washington and Richmond, not more than 100 miles apart, and connected by two lines of railway. Washington was only separated from Virginia by the Potomac, which is there a magnificent river, nearly a mile wide. Thirty miles higher up it is fordable.

5. The Shenandoah Valley, bounded east and west by high mountains, exceedingly fertile, and the great corn-growing district of Virginia. Not only did it supply the rest of the State, but it afforded a covered approach into Maryland, threatening the Federal capital.

6. The divided opinions of the border States, Missouri, Kentucky, and Maryland; and the very strong feeling in the north-western corner of Virginia in favour of the Union.

The first step the Northerners decided on was to blockade the Southern ports. The North had nearly all the vessels of the navy at its command; very few of the crews had joined the Confederacy, and it was thus possible to prevent supplies of any kind reaching the South from Europe. As the South was dependent for almost everything, except bread, meat, sugar, and tobacco, on other nations, the blockade was a most effective weapon against her. To starve her into submission did not seem difficult. She had no manufactures, except a few iron-foundries; no wool or cloth; no tanneries; no powder factories, no gun factories; almost all the railway workshops were in the North;

there was very little salt in her stores, and no tea or coffee. In fact, almost every single necessary of existence came from abroad, and had it not been that the arsenals within her territory were well supplied, and that her victories in Virginia provided her troops with equipment captured from the enemy, it is difficult to see how she could have carried on the war at all. As it was, the dearth of material resources always hampered her generals, as may be imagined when I state that they appear to have often depended for fresh supplies of ammunition on what they could take from the enemy.

The next step was to occupy north-west Virginia, and to deprive the Confederacy of this point of vantage. This was done without much difficulty, and the South was never able to reconquer it.

After the blockade had been established, and north-west Virginia occupied, the military policy of the Federals had two objectives.

1. In the east, the capture of Richmond.
2. In the west, the occupation of the Mississippi Valley.

Before the latter could be accomplished, the border States of Missouri and Kentucky had to be secured. These States were important to the Confederates as recruiting grounds, and they fought hard to retain them. But eventually the North proved superior. The border States were lost; and in July 1863, by the capture of Vicksburg, the great fortress of the Mississippi, General Grant made the river free to the Federal gunboats from New Orleans upwards, and thus cut the Confederacy in two.

During the third year of the war, July 1863 to July 1864, the Federals in the west were occupied in securing the State of Tennessee, and in pushing forward towards the lines of railway which connected the States of Georgia and Alabama with Richmond. Their progress was slow, and they met with stubborn resistance at every point.

Meanwhile, in the east, during these three years, the North had won no important advantage whatever. They had sent, at intervals, no less than five commanders into Virginia, with the purpose of capturing Richmond, but their armies had never won a single victory on Southern soil.

Twice had the Confederates, under Lee, crossed the Potomac; the first time into Maryland, in order to get recruits; the second time they had advanced into Pennsylvania. On both occasions Lee was compelled to retire; and in July 1863, the same month and almost on the same day that Vicksburg fell in the west, he was defeated at Gettysburg, in Pennsylvania, by General Meade.

Still, when the fourth year of the war opened, the Federals were very little nearer Richmond than they had been at the very outset. The capture of the chief city of the South and the destruction of her armies seemed as far off as ever.

In April 1864, the Northern people were scarcely hopeful. They saw no signs as yet of the end, and it seemed as if the frightful expenditure of life and money might drag itself on for years and years. But early in 1864 Grant had been appointed Commander-in-Chief of the Federal armies, and President Lincoln not only refrained from interference with his strategy, but gave him most loyal support.

Grant was a man of iron will and indefatigable energy, and he infused something of his own spirit into the operations of the Northern armies. His strategical conceptions, too, were broad and sound. Before he took over the chief command the Federal forces in the east and west had been entirely independent of each other; they had never worked in combination, and the Confederates, possessing the interior lines, had been able to transfer troops from one quarter of the theatre of war to the other without impediment. The Southern forces were divided into two main armies, one in Virginia, the other in the west, and Grant determined, with his superior numbers, to give these armies no respite, and to prevent the one from reinforcing the other.

The western operations were entrusted to General Sherman. The Commander-in-Chief accompanied the army moving against Richmond. As to Sherman's campaign, I need only say that it was completely successful, and had for its results the destruction of the army opposed to him, and a march across Georgia from Atlanta to Savannah, for the second time cutting the Confederacy in two, destroying the Southern arsenals and magazines, and isolating Virginia and North Carolina from the States on the south coast. Savannah was taken in December 1864.

But, although successful in Georgia, the Federals in Virginia, opposed by Lee and the finest of the Confederate armies, an army small in number but composed of veteran soldiers inspirited by many victories, met with the most determined opposition.

The first week in May Grant set out with 130,000 men to crush Lee's 60,000 and to capture Richmond. For a whole month the two armies fought day after day, the Federals dashing fiercely at the Confederate lines, recoiling with fearful losses, and then moving off to try to turn their enemy's flank. But no sooner was the Northern army set in motion than Lee moved too, and whenever Grant turned in the direction of Richmond he found his watchful antagonist still barring the way. At length after fifty days' marching and fighting, Grant found himself with the Confederate army between him and the Southern capital, holding the famous lines of Petersburg. He had lost in battle since the campaign commenced nearly 70,000 men, the Confederates not more than 25,000. But the Federal Government continued to pour in reinforcements, and his numbers were still almost twice as large as those of his opponent. But he had had enough of attacking the Southern breastworks; and, it is said, so appalled were the Northern people at the awful slaughter of their soldiers and so hopeless of success, that the Confederates were never so near to independence as in August 1864.

Grant now determined to lay siege to Petersburg, and to starve his enemy out. And indeed it seemed an easy task. If the war lay heavy on the North, it lay far heavier on the South. There were no more men to fill the ranks of her armies. The greater part of the country was exhausted by the march of the invaders. Old men and boys, unfit for service, were called upon to take their places at the front. As Grant himself said, 'the Confederacy was robbing the cradle and the grave to fill the ranks.' Of the sufferings in Richmond during the long siege of eight months it is pitiable to speak. The soldiers themselves were badly fed. The work at the front, with their inferior numbers, was unceasing and exhausting, and yet they bore it without complaint. But in the great city behind, in the hospitals, and in the homes of those whom the war had made widowed and fatherless, want and famine bore a far more terrible aspect.

And yet there were none who murmured. Whilst Lee and his army still held their ground that indomitable people never abandoned hope.

But at length the end came. Richmond was cut off almost on every side. Sickness and starvation had reduced the army to 40,000 men, and Lee was compelled to abandon the lines he had so long defended. He broke away; but it was too late. The net closed round him, and at Appomattox Court House, some seventy miles west of Richmond, the army of the Confederacy surrendered on April 9, 1865. The great war was over and the Union was restored.

Such is a very bare sketch of the salient points of the military operations.

Part II. The Strategy and Tactics of the Belligerents

I have already discussed the strategy of the American War in so far as it was affected by geographical and political considerations; I have now to deal with the actual strategical conceptions and operations of either side. As regards the main principle on which they acted, it has been said that the two belligerents fell naturally into their respective rôles. The North, intent on crushing out rebellion, was the invader; whilst the South, as Colonel Chesney writes in one of his essays on the war, as the weaker party outnumbered by nearly three to one, was compelled to stand on the defensive. Now, despite this very high authority, I cannot help thinking that the principle laid down, like almost every other military maxim, may be more honoured in the breach than in the observance. We know the stereotyped answer to all tactical problems, 'it depends on the nature of the ground.' That answer, vague as it is, is often the best that one can give. It implies that tactics are subject to no rule of thumb; and the same applies to strategy in general, and to the maxim we are speaking of in particular. There is no compulsion about it. The possibility of the weaker party assuming the rôle of invader depends not on the relative numbers of the two armies, but upon their *moral*, on their condition of readiness, and, above all, on the possibility of meeting the enemy in detail. Napoleon, whenever he could seize the initiative, never hesitated to throw himself into hostile territory, even when he was inferior in strength to the mass of the opposing forces; and it is remarkable that General Stonewall Jackson, certainly one of the greatest of American generals, constantly advocated the invasion of the North. But in the councils of the South political expediency over-rode military considerations. Defence not defiance was the motto of the young Republic; and her rulers, always trusting that sooner or later the European Powers would intervene in her favour, preferred that the Confederacy should pose as a State defending her liberties rather than as one seeking them aggressively. Twice only did General Lee, with the finest army of the South, cross the border and advance into Northern territory. On the second occasion he was met and defeated by Meade, at Gettysburg, in Pennsylvania, north of Washington; but to anyone who reads the history of the war, and realises the apprehension, the unreadiness, and the military weakness of the Northern States at the time the battle was fought, the truth of the saying that at Gettysburg the South was 'within a stone's throw of independence' is no less manifest than the wisdom, under the conditions, of an offensive policy.

But, preferring the defensive as they did, the Confederates made good use of their opportunities. Two points are remarkable. The main armies, one in Virginia and one in the west, were, generally speaking, always maintained at the greatest possible strength; strategical points, which lay outside the reach of these armies, were garrisoned by the very smallest force compatible with security. The principle was recognised that such points usually stand or fall with the success or failure of the larger operations. However, there was one remarkable and fatal exception. After the fall of Vicksburg no less than 55,000 men were retained in Texas and Louisiana, the trans-Mississippi States, and this at a time when the main armies of the South, for want of reinforcements, were absolutely unable to assume the offensive. Owing to the loss of the Mississippi these States were useless to the Confederacy. Fifty-five thousand men, who would probably have turned the scale elsewhere, were thus injudiciously employed in guarding unprofitable territory. It is only fair, however, to notice that there seems to have been a certain reluctance amongst a portion of the troops to serve outside their own States. The second point is the advantage afforded by the possession of interior lines. The Federal armies, invading the South from the north-west and north-east, were more than 1,000 miles apart; and when, after the second year, they had secured the border States and the Mississippi, they practically surrounded the enormous territory which the Confederates still possessed. Within this huge half-circle the Southern generals were free to move their troops as they wished. They used their freedom to some purpose. The point most actively threatened was again and again reinforced from the other quarter of the theatre of war. Thus, in 1863, after Gettysburg, 20,000 of Lee's army, under Longstreet, one of his best generals, were sent to the west, and enabled the army in that section to gain the important victory of Chickamauga, which for several months completely paralysed the Federal advance into Georgia.

At the same time it must be said that this constant and effective shifting of strength from one wing to the other was made feasible by the errors of the Federals. Their two main armies of the west and east worked on wholly independent lines. Until Grant took command in 1864, they never operated in combination. Whilst one was moving forward the other was resting or preparing for a fresh advance; and this disjointed state of things permitted their enemy to reinforce the threatened point at his leisure. Grant initiated a new policy. He pressed his opponents at every point simultaneously. Relying on his superior numbers he neutralised all the Southern advantages of interior lines. It may be argued that this strategy entailed a useless waste of life; that the better plan would have been to hold the enemy on one wing, and to attack him in force upon the other. But here we must remember the enormous extent of the theatre of war. It was easy enough for the Southern armies to get across the Confederacy in a very short time, and, by destroying the railroads, to make pursuit hopeless. This was prevented by Grant's energy in pushing the attack at every point.

The Federal strategy of the last year of the war, with Grant in command and Sherman his lieutenant, stands out in marked relief to the disjointed, partial, and complicated operations of the previous years. The plans of campaign evolved during the first phase of the war were ingenious in the extreme. Simplicity was despised. The great idea was to surround the enemy, to cut off all his communications, and to attack him in front, flanks and rear, at one and the same time. Unfortunately this conception made it necessary to break up the invading army into several columns, and the enemy, using his interior lines, had little difficulty in spoiling the whole plan. He either defeated each column in succession or, by crushing one of them, compelled the others to fall back. The second invasion of Virginia, in 1862, was carried out by no less than four different armies, all converging on Richmond, and numbering all told about 200,000 men. The Confederates had but 100,000, but the brilliant strategy of Lee, backed up by the marvellous energy of Jackson, cleared Virginia of invaders within three months. This tendency to discard simplicity in favour of complication appears in the tactics of the Federals as well as in their strategy. Commanders were always trying to imitate Napoleon, forgetting that intricate manoeuvres require a well-trained staff and well-drilled troops; and it is noticeable that the least experienced leaders were generally the most eager to attempt involved movements. Grant seems to have been the first to recognise that, as Moltke puts it, the true objective of a campaign is the defeat of the enemy's main army, although he may be said to have erred on the side of simplicity, and too many of his battles took the shape of frontal attacks against an entrenched enemy. General Sheridan's summing up of the handling of the Army of the Potomac, as the army of the east was called, before Grant took command, is to the point. 'The army,' he says, 'was all right; the trouble was that the commanders never went out to lick anybody, but always thought first of keeping from getting licked.' Grant, like Moltke, was always ready to try conclusions.

Perhaps the most interesting strategical question is that connected with bases of operations and lines of communication. Grant was the first to perceive that in a comparatively fertile country it was possible to subsist an army without magazines; and he was able to invest Vicksburg, the Mississippi fortress, by cutting loose from his base, marching completely round the place, defeating the troops that opposed him, and then establishing a new line of communication. In his famous march to the sea Sherman did the same thing. In September he found himself at Atlanta with a Confederate army, inferior in numbers, in front of him, and in October this army passed round his flank and struck his line of communications in rear. But his magazines, depots, and the important bridges were fortified and well garrisoned; the border States, Kentucky and Tennessee, were strongly held; and so on November 15, cutting loose from his communications, he started on his march of 300 miles across Georgia, entering Savannah on December 21. The Confederate army of the west, which he had left in his rear, was heavily defeated at Nashville on December 15 and 16.

I may add that the command of the sea and of the great rivers both in the west and east greatly assisted the Federal generals in their operations, as they assisted General Ross of Bladensburg, in that remarkable campaign which resulted in the capture of Washington by an English army.

In 1864, Grant, with an army 130,000 strong, moved southward against Richmond through Virginia, always keeping his left within reach of the navigable rivers and estuaries which intersect the eastern portion of that State in a direction parallel to the line of march. The district through which he moved was completely exhausted, and he was compelled to rely on his magazines. In fifty days he changed his bases and line of communications no less than four times; a fact which speaks volumes for the efficiency of the Federal departments of supply and transport.

In comparing the broad principles of the strategy of the Federals with those followed by Moltke in 1870, we are at once struck with the complication, the vagueness, and the weakness of the one, as compared with the simplicity, the strength, and the concentrated energy of the other. In 1870 we find a vast army, divided into two groups, disdaining every object except that of concentrating every single available gun, sabre, and bayonet against the main forces of the enemy. Everyone is aware that Moltke's plan of campaign, seemingly so simple, had been most carefully worked out in the winter months of 1867–68. The Federals, on the other hand, not anticipating war, had no such opportunity of thinking out at their leisure the proper line to be followed, and the result was that for the first three years they made but little progress. Now the Federal generals were, as a rule, men of strong common sense, and it is often urged that strategy is merely a question of common sense, but in 1870 we have one of the most earnest students spending four months in evolving a plan of campaign which proved completely successful, and in 1861, '62, and '63, men of undoubted ability, producing and acting upon conceptions of which the most ordinary Sandhurst cadet is able to point out the shortcomings. Common sense made a most conspicuous failure. It is true that the Federal generals were much hampered by the President and his advisers, who never ceased, until the coming of Grant, to interfere with the military operations; but the fact that the ideas of these civilian councillors were almost invariably unsound goes to prove the proposition that for judicious strategy something more is needed than mere natural intelligence. In General Grant's Memoirs is an anecdote much to the point. When he first took command, he had an interview with the President, Abraham Lincoln. Now Lincoln was undoubtedly one of the very ablest men that America ever produced; he had given advice to every general-in-chief, had received every report, and had naturally followed the course of the war with the most intense interest. Yet mark the following;—'In our interview,' says Grant, 'the President told me he didn't want to know what I proposed to do. But he submitted a plan of campaign of his own which he wanted me to hear and then do as I pleased about it. He brought out a map of Virginia . . . and pointed out on that map two streams which empty into the Potomac, and suggested that the army might be moved on boats and landed between the mouths of these streams. We would then have the Potomac to bring our supplies, and the tributaries would protect our flanks while we moved out. I listened respectfully, but did not suggest that the same streams would protect Lee's flanks while he was shutting us up.'

I think that when we compare the strategy of the American war with that of 1870 we realise the truth of Napoleon's saying: 'Read and meditate on the wars of the greatest captains. This is the only way of learning the science of war.'

Now, as to the tactics of the three arms.

To take the Artillery first. In the first year of the war we find, as we should naturally expect, knowing that the batteries had never had opportunities of working together, that in battle, whether on the defensive or offensive, their action was entirely independent. In 1862, however, came a change. The first symptom was seen at the battle of Malvern Hill, where the Federal army, retreating from before Richmond after a crushing defeat, fought a most successful rearguard action. Its success was due not so much to the strength of the position as to the fact that the chief of artillery had massed nearly 300 guns to meet the attack of the Confederates. This principle of massing guns gradually worked its way to the front, and the last great charge of the Confederates at Gettysburg, in July 1863, was preceded by an artillery duel for nearly two hours, with 137 guns on one side and about 90 on the other. The ground on which this battle was fought, however, was fairly open. In the forests of Virginia space for the deployment of more than six or seven batteries at most was seldom to be found; in fact there was so little opportunity for its employment, and it was so liable to capture, that, after a few days' campaigning in the Wilderness in 1864, General Grant sent back a large portion of his artillery. Naturally, in such a country, surprise played a most important part in the operations, and one of the Federal generals, Hazen, in his memoirs, speaks somewhat contemptuously of the 'old custom of advertising one's intentions by a cannonade.' He is of course referring to fighting in a very close and intersected country.

Shrapnel was little used in the war, and the guns were far from possessing the killing power of those of the present day; it is, therefore, scarcely worth while speaking at length on the effects of artillery fire on the troops. Generally speaking, the effect, whether moral or physical, was very small. Like all raw troops, in the first year of the war the men appear to have dreaded the artillery a good deal: but when they found that 'masked batteries,' a great bugbear in the earlier days, were very seldom met with, and that the losses inflicted by the artillery were out of all proportion to the noise, contempt seems to have taken the place of apprehension. At all events, neither infantry nor mounted riflemen had the slightest hesitation in charging artillery, and I doubt if any troops ever faced guns with less perturbation of spirit than the Americans.

At Fredericksburg, in December 1862, the Federal army was on one side of the river Rappahannock, on commanding ground; Lee's army on the other, well out of range, but holding the little town of Fredericksburg, on the bank of the stream, with a small brigade. Before crossing the river, the Federal commander determined to clear the town, and bombarded it for nearly an hour with some fifty or sixty guns, including several 20-pounders. 'Although the effect on the buildings was appalling; although flames broke out in many places and the streets were furrowed with round shot, the defenders not only suffered very little loss, but at the very height of the cannonade easily repelled an attempt on the part of the Federals to cross the river.' The stream was eventually crossed in boats, and the Confederate brigade driven out at the point of the bayonet.

Daring, characteristic of all arms, was very conspicuous in this branch of the service. In the Mexican War of 1846–47, the field artillery had done excellent service, always pushing forward with the fighting line. It had brilliant traditions, and one of the marked features of the Civil War is the almost reckless fashion in which the batteries assisted the attack. They were often to be found in line with the most advanced skirmishers, and rendered the infantry the most effective support. No false shame of losing guns ever kept the battery commanders back when they could do good work at the front, and the greater part of their fighting was done at canister range.

The Southern artillery was much inferior in material; the fuzes were very bad and the ammunition indifferent; but, on the whole, it did remarkably good work. This was due to a more judicious organisation. The artillery officers in General Lee's army were given a much freer hand than in the North. The chief of artillery in each army corps advised his chief on all matters appertaining to his own arm, and all tactical details were left to him and the officers under him. Four batteries formed a battalion, generally attached to an infantry division, but not permanently to any one division in particular, and these battalions were very seldom split up.

In the Northern army, a varying number of batteries were attached to each infantry division; but there was always a disposition to allow the divisional commanders to use their batteries as if they were independent commands, and not as if they constituted only a section of a unit. Chiefs of artillery were considered useless; there were no competent staffs; and, generally speaking, there was an absence of concentrated effort on the part of this arm which greatly minimised its effect. The divisional commanders were accustomed to use their guns without reference to the artillery officers, and hence, as one of the artillery generals writes, 'idle cannonades were the besetting sin of some of our commanders.'

There are two points connected with the artillery duel which I may notice in passing.

The first has reference to artillery on the offensive.

The Americans appear, like the Germans at Gravelotte and elsewhere, to have generally limited the action of their batteries to merely silencing the enemy's guns; and the preliminary bombardment had often very little effect on the issue of the fight.

Thus, at Gettysburg, of which I have already spoken, the Federal artillery commander, after maintaining a rather unequal contest for nearly a couple of hours, ordered his guns to cease fire. His opponents, as he implied in an account he wrote of the battle, had sufficiently advertised their intentions; and he simply ceased fire to save ammunition for the infantry attack which he knew must follow. The Confederates, believing that they had silenced him altogether, let loose their infantry, on which seventy Federal guns opened at short range with terrible effect. In fact, it was St. Privat anticipated.

The generals on both sides took very good care to keep their infantry either well under cover, or well to the rear, while the artillery duel was going on. During the bombardment, preliminary to the Federal attack on Lee's position at Fredericksburg, I believe that the Confederate front was manned by no more than half a dozen battalions at most. The main army of 80,000 men was hidden in ravines and woods, well out of range. In one of the Wilderness battles, forty Federal guns were engaged for a long time bombarding a line of earthworks from which the garrison had been withdrawn to the shelter of the neighbouring forest.

At Gettysburg the Federal infantry were not withdrawn. They lay in open order, behind slight entrenchments and stone walls. They were not very far in front of their own batteries, and the latter were the Confederate objective. I have looked through the reports in the official records sent in by the infantry regiments. All of them speak of the bombardment, but none of them appear to have lost more than one or two men from the fire of the guns. When the Confederate infantry advanced, these regiments were unshaken and perfectly ready to do their share of the business.

In fact, the attacking artillery had only carried out the first part of the bombardment; it had silenced the opposing batteries, but it had done nothing whatever towards destroying or demoralising the opposing infantry.

On other occasions the infantry assisted the artillery in the bombardment, and here the infantry of the defence were compelled to show themselves. They could not be withdrawn from their earthworks with an infantry force watching its opportunity not many hundred yards to the front, and gradually creeping up under fire of its own guns. They were obliged to join in the action, and directly they exposed themselves above their entrenchments the artillery took them for its target.

At the battle of Nashville, December 1864, where the Confederate army of the west was finally defeated, the day was won by a smart stroke of combined tactics.

A hill which formed part of the Southern line was strengthened by an earthwork. The Federals massed guns against this point, and sent a brigade across the valley to storm it. 'The fire of these guns,' says the Confederate commander, 'prevented our men from raising their heads above the earthworks, and the enemy's infantry made a sudden and gallant charge up to and over our entrenchments. Our line, thus pierced, gave way; soon after it broke at all points, and I beheld, for the first and only time, a Confederate army abandoning the field in confusion.'

As to the infantry, the battalions on either side, organised in ten companies, used a drill which was more French than English; all movements were very quickly carried through, and much use was made of skirmishers to cover the advance of the line or column. The usual formation for attack was in line, with either two companies per battalion or a battalion per brigade deployed as skirmishers. Attacks in close column were infrequent; and the advance was generally made in successive lines, as was advocated by Skobeleff. In fact, Skobeleff, who, according to Archibald Forbes, was an earnest student of the American War, seems to have adopted many of his ideas from the practice of the great American generals.

There are not many points of peculiar interest about the infantry tactics.

One of the Federal divisional leaders, General Hazen, complains that there was a singular lack of tactical manoeuvring in the war. Many battles, he says, were little more than the posting of lines to give or receive the attack. The men then fought the matter out in their tracks, and the affair ended with a disorderly retreat or a broken and ineffective pursuit. 'This was in part,' he writes, 'due to the too loose moulding of the regiments by drill and discipline, . . . but very largely to the lack of a staff clearly comprehending the situation and needs of the moment.'

It is perhaps more probable that this lack of manoeuvring was due to long-range firearms. General Hazen seems to me to be comparing the battles of Lee and Grant with those of Frederick and Napoleon; for the same lack of manoeuvring under fire—for this is what he refers to—was just as apparent in 1870, and is a necessary evil of modern fighting. The American advance was made in what were literally successive lines of skirmishers. The men opened out under fire, and abandoned touch of their own accord. It has been said that the Germans, in 1870, adopted extended order because they held the very curious belief that therein lay the royal road to victory. I hold, myself, a very contrary view. I believe that the Germans extended their men because they knew it was impossible to get them to advance in any other formation under the stress of modern fire. And in this opinion I think American soldiers will be found to agree. At all events a veteran of the Civil War, who commanded a famous volunteer regiment, when I asked him whether men could be got to advance shoulder to shoulder in close order under the fire of the breech-loader, gave a most decided negative. 'No,' he said, 'God don't make men who could stand that.'

One of their great generals thus speaks of the Confederate attack. 'Whoever saw a Confederate line advancing that was not as crooked as a ram's horn? Each ragged rebel yelling on his own hook, and aligning on himself!'

If the attack by successive lines of skirmishers was invented by the Americans, so also was the advance by means of successive rushes. 'The troops,' says the officer just referred to, 'invented the attack by rushes, that is, they fell into the habit of making their attacks that way, because it was the only way to work sensibly.'

The attack of large bodies underwent a marked development during the war. It is curious to find an experienced leader like Sherman, at Bull Run, the first great battle, sending the battalions of his brigade into action successively; when one was beaten another took its place. But Sherman, like many of the others, had to buy his experience in the field. In the earlier period, and indeed generally speaking right through, the traditional English formation of skirmishers, followed by three lines, seems to have been universal; but in the third year there was a tendency to mass troops on a great depth for the assault of the tactical objective. At Gettysburg, after the great artillery duel of which I have already spoken, Lee put in 15,000 men to breach the Federal centre, and, but for some misunderstanding, they would have been followed by 15,000 more. At Chickamauga, two months later, Longstreet formed seven brigades, in column of brigades at half-distance, and in this formation made a successful breach of the Federal lines. At Spotsylvania, in May 1864, Grant massed no less than 30,000 men for the assault of what was afterwards called the 'Bloody Angle,' so fierce was the fighting and so terrible the slaughter round it. The centre of this attack was formed of two divisions in line, and two in column in rear. It was but partially successful. The supports mingled with the first line as they stormed the entrenchments; there was another strong line of earthworks in rear, and here the Federals were roughly checked.

At Chattanooga, in November 1863, Grant carried the centre

of the Confederate position, a ridge 500 feet high, with four divisions disposed in three lines.

At Chancellorsville, May 1863, Jackson's famous flank attack, which rolled up the right wing of the Federal army, was made by 25,000 men, drawn up by divisions in three lines, and covered by skirmishers.

It is important to note that the attacks in which each line was formed of a single division appear to have been far more productive of confusion, and were never so thoroughly successful, as those where each division was drawn up in three lines, as at Chattanooga.

I can do no more than refer very briefly indeed to Lee's great flank attacks, made with every man that he could spare, imitating Frederick the Great, and anticipating the decisive movement of the 12th Corps upon St. Privat. Tactically speaking they were the most brilliant manoeuvres of the war.

I have already spoken of the very slight control that the regimental officers exercised over their men when the bullets began to fly. This absence of fire discipline greatly increased the difficulty of supplying ammunition.

'The complaint,' writes General Hazen, 'out of ammunition,' used to be heard from regimental commanders fifty times during a great battle. This was often due 'to want of control over the fire owing to poor drill.' Here he is referring to the infantry, who were armed with muzzle-loaders; and it is an interesting fact that General Lee, owing to the same difficulty of control, was averse from arming his infantry with breech-loaders.

The ammunition was kept in the battalion carts, and the packets carried to the firing line in bags, but the supply, or rather the means of bringing it up, were very often unequal to the demand.

I think that these are circumstances well worth the closest attention of those who may have to deal with unprofessional troops; and that the more we read of the American War, the more we realise the value of steady drill and strict discipline. At one time many of the Federal soldiers in the west threw away their bayonets; and in Sherman's march to the sea the men got rid of their knapsacks, finding it more comfortable to march with their necessaries rolled up in the blankets that were slung round their shoulders.

This very rough description of the American artillery and infantry shows that their tactics differed little, if at all, from those now in vogue in Europe; but in the tactics adopted by their mounted riflemen we come to what was practically a new feature in modern war. This new departure was due principally to the nature of the country; the mounted arm on either side, at all events in Virginia, was in no sense of the word mounted infantry, that is, soldiers who use the horse merely as a means of locomotion, as transport for their rifles, instead of as the principal and most direct means of defeating their enemy.

The truth is, I believe, that the American mounted regiments, after the first two years, when they had become sufficiently trained, preferred to fight on horseback rather than on foot. But they were accustomed always to adapt their tactics to the ground. If the ground was unsuitable for mounted work they converted themselves into infantry. If they engaged infantry, they fought that infantry with its own weapons so long as it gave them no opening for a charge. And, as a matter of fact, the ground generally compelled them to fight dismounted. The best way, I think, of opening a discussion on the merits and value of this force is to put the question:—Were these mounted riflemen efficient both as infantry and as cavalry? This, I take it, is what we all want to get at. Do the records of the mounted riflemen of America assist us to decide the much-vexed point whether cavalry can be so trained as to work well on foot without impairing their efficiency when mounted? When I use the term 'much-vexed question,' I do not wish to be misunderstood. It has been settled in England by the action of our own authorities in establishing training schools for mounted infantry.[9] But other nations refuse to be convinced. I can, of course, do no more than bring forward certain facts and offer certain suggestions; and I must preface my remarks by saying that, owing to the different conditions of warfare in America from those

9. [These schools, established by Major (later Major General) Sir Edward Hutton with the enthusiastic support of Lord Wolseley, were situated at Aldershot and the Curragh (A. J. Godley, "Mounted Infantry Training at Home," *Cavalry Journal*, I [January, 1906], 52–55).]

that obtain in Europe, and the meagre records of their mounted branch, the evidence I shall produce will possibly be insufficient to warrant a verdict either way.

I have already alluded to the efficiency of the American men on the outpost line. But there were two other tactical operations in which they shone even more conspicuously. One is the 'raid'; those extraordinary enterprises which did so much harm to the enemy's communications, and so completely thwarted and disordered his manoeuvres. The other is the delaying power possessed by the mounted arm; the manner in which the cavalry and horse artillery alone were able to check for many hours the advance of the enemy's infantry or artillery, or to hold that infantry and artillery fast until reinforcements arrived. Every soldier knows that the American mounted riflemen possessed a most remarkable strategical independence; and, as Sir George Chesney, with these riflemen in his mind, long ago asserted, '30,000 such horsemen would, if handled boldly, wholly cripple and confound an opposing army of 300,000! Riding to and fro in rear of an army, intercepting its communications, cutting off its supplies, destroying its reserve ammunition and material, such a force would undoubtedly create panic and confusion far and wide.'[10] That all cavalry should possess this measure of strategical independence we are probably all agreed. The question is, can it be done? The reply is, certainly, if the mounted arm can fight equally well mounted and on foot; if it combine, as did the American horse, shock and fire-action.

Now, I will try to explain, in as few words as possible, the standard of efficiency reached by Stuart's and Sheridan's mounted men. First as infantry. I do not think that anyone

---

10. [Sir George Chesney, brother of Charles Cornwall Chesney, was a well-known soldier and military writer. His most famous work is "The Battle of Dorking: Reminiscences of a Volunteer" (*Blackwood's Edinburgh Magazine*, CIX [May, 1871], 539–72), which purported to be the story of a successful German invasion of England and which was written to urge action in improving the efficiency of the Volunteer forces. He was one of the first to advocate the conversion of the Yeomanry into mounted infantry (see "A True Reformer," *ibid.*, CXIII [April, 1873], 474). The Chesney Gold Medal, to be awarded "to the author of original literary work treating with naval or military science and literature which has a bearing on the welfare of the British Empire," was instituted in 1899 in memory of Sir George. Curiously enough, the first to receive the medal was an American, Captain A. T. Mahan (information obtained in correspondence with Captain B. H. Liddell Hart).]

dare assert that their best mounted regiments, when fighting on foot, were anything like so efficient as the ordinary infantry. Read Sheridan's account of the battles of Five Forks and Sailor's Creek, where his command gained its brightest laurels, and you will observe a note of triumph when he writes that his dismounted cavalry were able to hold their own against the Confederate infantry. These battles occurred at the very end of the war, and I believe that it was not till that time, four years after the war began, that the cavalry fancied themselves anything like a match for the infantry. And, at the same time, we must always bear in mind that the cavalry were armed with breech-loading and repeating carbines, the opposing infantry with muzzle-loaders. This last is a most important point, and was, of course, all in favour of the cavalry. 'The difference,' writes Stuart's Adjutant-General, 'between a Spencer carbine and an Enfield rifle is by no means a mere matter of sentiment.' This evidence refers to the eastern theatre of the war. A single extract will show, I think, that at the same period the infantry in the west had very little dread of the trooper on foot. Writing of the last great battle in the west, Nashville, a Federal staff officer, describing the advance of the Confederates, writes, 'Bradley was assailed by a force which the men declared fought too well for dismounted cavalry.' This shows the estimation they were held in in the west, and I think we are justified in believing that the cavalry, notwithstanding their superiority of armament, were only fair infantry.

Secondly, were they good cavalry? Let us divide the battle duties of cavalry into the attack on infantry and artillery, and the attack on cavalry. Now it seems to me that these two duties require very different qualifications, and that the latter is by far the more difficult. Indifferent cavalry, so long as the men ride well and their hearts are in the right place, can charge successfully even good infantry and artillery, if surprised or demoralised; but the same troops, were they to meet good European cavalry on a fair field, would be nowhere. Now the American cavalry had never much hesitation in charging guns; and in the last year of the war, when Sheridan came to the front, they were just as capable of charging infantry as either the French or Germans in 1870. I believe that there were many brigades in both the Federal and Confederate armies who would have charged just as gallantly, and possibly just as far, as did von Bredow at Mars-la-Tour; but whether, in a country far more open than their own, they could have met the German cavalry of that date with any hope of success, or whether they could have done all that the Germans of to-day anticipate may be done by enormous masses skilfully manoeuvred, is a very different question, the solution of which is beset by many difficulties. In the first place, we are all aware that it takes a long time to train cavalry to manoeuvre in mass with speed and cohesion; and, also, that without manoeuvring capacity you can scarcely hope for success against hostile cavalry thoroughly well trained; nor, without high manoeuvring capacity, would enormous masses achieve against infantry the results anticipated in future wars by the Germans.

Now I ask whether it is likely that either the Federals or Confederates, beginning with men and horses absolutely untrained (except in so far that the Southerners all rode well), and with only a few senior officers, and fewer non-commissioned officers, who knew anything of their work, should have been able to acquire, during incessant active service, great manoeuvring capacity. We know that the thorough training of the horses has much to do with the efficiency of the German cavalry. It is impossible that the Americans, who had no establishments of trained horses to fall back upon, no depots at which to train the remounts, and who had to supply casualties with horses unseasoned and impressed straight from the farm, should have been able to approach European cavalry in mechanical perfection of movement. And yet from this mechanical perfection come rapidity of manoeuvre and cohesion. The two qualities are absolutely essential to success in a cavalry engagement. Again, there was want of discipline. To quote a Lieutenant-General of the Confederate army: 'The difficulty of converting raw men into soldiers is enhanced manifold when they are mounted. Both man and horse require training. . . . There was but little time, and it may be said less disposition, to establish camps of instruction. Living on horseback, fearless and dashing, the men of the South afforded the best possible material for cavalry. They had every quality but discipline. . . . Assuredly our cavalry rendered much excellent service, especially when dismounted and fighting as infantry. Able officers, such as Stuart, Hampton, &c. &c. developed much talent for war; but their achievement, however distinguished, fell far below the standard that would have been reached had not want of discipline impaired their efforts and those of their men.'

However, these are but opinions; and I will now give as a practical illustration, a sketch of the most famous cavalry battle of the war, that of Brandy Station, fought in Virginia on June 9, 1863. I may say, first of all, that before this engagement there had been plenty of hand-to-hand fighting and cavalry charges, but the charges were made in column of sections down the roads.

On June 9, 1863, the Confederate cavalry, under Stuart, was stationed near Brandy Station, in Virginia. The Federals were on the other side of the river Rappahannock, the numerous fords being held by the Confederate pickets. Stuart had ordered his division, of seven brigades, about 10,000 strong, to march at an early hour; but at the very earliest dawn, the Federal cavalry, under General Pleasonton, consisting of three small cavalry divisions with infantry supports, and also about 10,000 strong, crossed the Rappahannock in two columns, with the intention of reconnoitring towards Culpeper Court House. The right column crossed at Beverly, the left at Kelly's Ford, about five and a half miles lower down the stream.

We will take the right column first. It had some difficulty in dislodging the Confederate picket and support, and here there was a good deal of hand-to-hand fighting on a narrow road. Eventually the Confederates, who formed part of General Jones' brigade, were pushed back to St. James' Church, where they found three of their brigades drawn up in position, dismounted, under cover of stone walls, forming the front line, with mounted regiments on the flanks. The Federals dismounted and attacked the left wing of this position; but they were repulsed, and charged, it is said, by cavalry. But whether there was any hand-to-hand fighting at this point there is no evidence to show. On the Federals falling back the Confederates advanced; and it seems that for several hours there was a great deal of skirmishing, relieved by a dashing charge of Federal cavalry.

MAP V

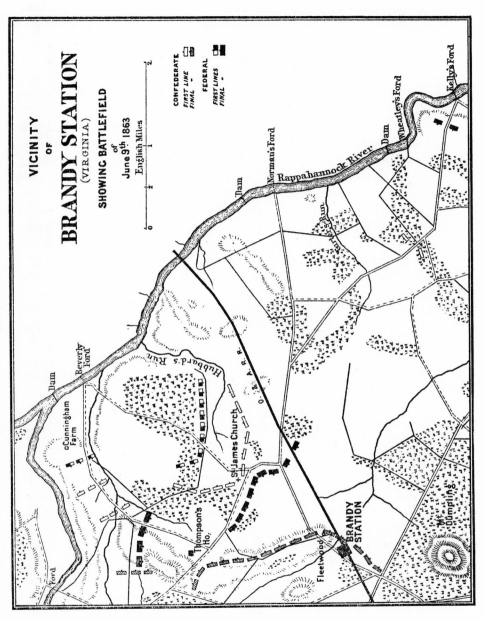

This was made by the 6th United States (regulars). 'It was made,' says an eyewitness, 'over a plateau fully 800 yards wide, and its objective point was the artillery at the church. Never rode troopers more gallantly than did those steady regulars, as, under a fire of shell and shrapnel and finally of canister, they dashed up to the very muzzles, then through and beyond our guns. Here they were simultaneously attacked from both flanks and the survivors driven back.'

Now for the left Federal column. It crossed the river without difficulty, and then divided into two columns, two divisions advancing on Brandy Station, the other on Culpeper. Eluding a Confederate brigade, which had come up to support the picket at the ford, the two columns moved forward. The first was soon seen by the Confederate scouts to be moving directly on Brandy Station, and when reported was visible from Fleetwood Hill and was actually in rear of the Confederate lines engaged beyond St. James' Church. Fleetwood Hill, although nothing more than a gentle undulation, commanded the whole of the neighbouring country. Stuart was at the front; but he had left his Adjutant-General on the hill, having selected it as his headquarters during the action, and this officer, who had a single howitzer with him, but no troops beyond a small escort, opened fire on the Federal column and sent an urgent report to Stuart. The Federals halted and their horse artillery came into action. Stuart, on receiving the message, sent back a couple of regiments from the centre of his line to Fleetwood Hill. 'The emergency was so pressing,' writes the Adjutant-General, 'that the leading regiment had no time to deploy. It reached the top of the hill just as the single piece of artillery was retiring. Not fifty yards below a Federal regiment was advancing in magnificent order, in column of squadrons. A hard gallop had enabled only the leading files of the 12th Virginia (a Confederate regiment) to reach the top of the hill, the rest stretching out behind in column of sections. With the true spirit of a forlorn hope the colonel and a few men dashed at the advancing Federals, but did not check their advance. The other Confederate regiment now came up, but so disordered by their rapid gallop that after the first shock they recoiled and retired to re-form.' This left the Federal regiment in possession of the hill; the two Confederate regiments, having re-formed, again charged and drove them back for a time, but eventually had to retire leaving the Federals masters of the situation. Two squadrons, passing round the west side of the hill, charged the Federal horse artillery, which had advanced to its foot. The cavalry escort was dispersed, but the gunners fought splendidly, and the Southerners were unable to carry off the guns. The officer commanding the Federal battery reports that he was 'surrounded by a squad of rebel cavalry, firing with carbine and pistol.' By this time Stuart and the greater portion of his force was on his way back to Fleetwood Hill. One regiment from the right of the line led the advance, and, moving towards Brandy Station, was ordered by Stuart to charge a Federal force, the main body of the column, near the Miller House. The enemy appear to have been surprised by this attack, and were retiring slowly at the time. The attack was successful, and the squadrons even rode through a section of artillery; but the Federals re-formed, charged the Confederate regiment with far superior numbers, and drove it back. The Confederate charge—the regiment was but 200 strong—was made in line. By this time, the whole of the right wing of Stuart's first line, consisting of four regiments, was retiring on Fleetwood Hill, the whole force in column of squadrons.

Two of his regiments appear to have moved straight on Fleetwood Hill, which was now in possession of the Federals. The colonel of the leading regiment reports: 'I immediately ordered the charge in close column of squadrons, and swept the hill clear of the enemy, he being scattered and entirely routed.' The regiment in first line, according to Stuart's Adjutant-General, used the sabre alone, but it does not appear that the opposing cavalry rode out to meet the charge.

The two remaining regiments diverged to the left, passed the eastern end of the hill, and encountered the enemy, who had not long before driven back the first Confederate regiment. 'This charge,' says an officer present, 'was as gallantly made and gallantly met as any the writer ever witnessed during nearly four years of active service. Taking into estimation the number of men who crossed sabres in this single charge (being nearly a brigade on both sides), it was by far the most important hand-to-hand contest between the cavalry of the two armies. As the blue and grey riders mixed in the smoke and dust, minutes seemed to elapse before its effect was determined. At last the intermixed and disorganised mass began to recede, and we saw that the field was won by the Confederates.'

After this the Federals abandoned this quarter of the field, and a portion of their force which held the railway station was driven out by the charge of a fresh Virginia regiment, sent in by Stuart. The enemy then fell back to join that portion of his force which still remained near St. James' Church, having re-formed without molestation on the ground from which he had originally advanced.

It is noticeable that Stuart, on falling back in the first instance to Fleetwood Hill, was not followed by the enemy with whom he had been hitherto engaged.

American writers attribute the failure of the Federals to follow him to the position held by the Confederate brigade about Cunningham Farm, within striking distance of the road by which the Federals had advanced from Beverly Ford.

But this Confederate brigade, with its right flank exposed, was withdrawn, without molestation, to the hills overlooking Thompson's House; and Stuart's line now extended along these hills, but with a gap between the extreme left and the river.

The efforts made by the Federals to penetrate through this gap and get round Stuart's rear led to some more cavalry fighting. A brisk dismounted skirmish was followed by the charge of two Federal regiments. This was met by the 9th Virginia, which seems to have broken the attack and driven its assailants back across a stone wall. The 9th was then attacked in flank by a fresh regiment, and was driven back in turn; but being reinforced by the 10th and 13th regiments, it again advanced, and the tide of battle was finally turned against the Federals. Whether the sabre was used in these charges or in what formation they were made does not appear. They are but little noticed by writers on the war. A Confederate brigade now came up from Oak Shade [Church] to fill the gap, and after a short dismounted action the Northerners retired across the Rappahannock. 'No serious effort,' says Stuart's Adjutant-General, 'was made to impede their withdrawal'; but we may remember that the Federals had a small brigade of infantry present. While the main bodies had been engaged at Fleetwood Hill, two Confederate regiments had cut in across the line of march of the Federal division, making for Culpeper, and occupied a strong position on thickly wooded rising ground just east of Stevensburg. The fighting was principally dismounted; but on one of his regiments being caught changing formation, when mounted, on a narrow road, and dispersed by a charge in column of sections, the officer commanding the Confederates withdrew to Mountain Run, covering the road to Brandy Station. The Federal division was almost immediately recalled across the Rappahannock.

The Confederates, out of a total of 10,300, lost 523 officers and men. The Federals, out of 10,980, lost 936, including 486 prisoners and 3 guns.

Owing to some misunderstanding, the Confederate brigade, near Kelly's Ford, remained in that position all the morning and took no part in the engagement.

The action, from the time the Federals crossed the river to when they recrossed it at Beverly Ford, seems to have lasted about eight hours.

Another important engagement was that at Gettysburg, where the two cavalries came together on the right flank of the Federal position. Here, again, the only important charge of the day was made by two small brigades, numbering probably not more than 800 men apiece, in column of squadrons. Met by a regiment, also in close column of squadrons, in front, and attacked by several small parties in flank, this charge was beaten back.

To show how very far removed the cavalry fighting was from European ideas, I may mention that the charge of a Virginian regiment, which a Northern writer, in the 'Century Magazine,' records as the most determined and vigorous he ever saw, was made against a stone wall, on the other side of which was a Federal regiment, and hand-to-hand fighting, naturally with the pistol and carbine, took place across this barrier.

These were the most important instances of cavalry fighting; and, in my very humble opinion, it does not appear that, as a mounted force, so far as shock-action goes, the American cavalry came near the European standard. To sum up, my impression is—I give it for what it is worth—that they could charge infantry when surprised or demoralised; that they fought well on foot, but were not equal to well-trained infantry; and that, as cavalry, they were deficient in manoeuvring power and in cohesion.[11]

It is true that, in October 1864, Sheridan's cavalry, in the more open district of the Shenandoah Valley, did extraordinary execution, and combined with the infantry in a manner which makes his victories models of tactical skill. But his enemy was far inferior in numbers; and whilst the infantry attacked them in front, the cavalry was free to manoeuvre at leisure against their flanks and rear. Unfortunately, no account of his campaign with which I am acquainted goes sufficiently into the details of the cavalry fighting. All considerations as to formations, pace, time, and distance are unnoticed.

However, as regards cavalry *versus* cavalry, I have given a sketch of what are considered the most important engagements of the war; and I must leave it to my readers to decide whether the action at Brandy Station, with its single charge in mass, and that in close column of squadrons, like Stuart's, at Gettysburg, indicate a capacity for manoeuvring or a knowledge of purely cavalry tactics such as would have fitted the American horseman to cope in the open with good cavalry on the European model. I may add that at the battle of Winchester, October 1864, where Sheridan's cavalry so much distinguished itself, the charges against the Confederate infantry were again made in close column of squadrons, and I think it is a fair presumption that this was the usual formation whenever the cavalry were employed mounted in mass.

At the same time, no troops could have been better adapted to the country over which they fought then the American mounted riflemen; no troops ever showed greater pluck; on the outposts they were exceedingly efficient; their strategical independence was great, and, as I have already said, on their own ground they would probably have defeated any European

11. [For a detailed comparison of Civil War cavalry and the German cavalry after 1870, see Major Justus Scheibert, *Der Bürgerkrieg* (Berlin, 1874), pp. 71–74. Scheibert rated the German cavalry superior to the American cavalry in most respects.]

cavalry of the period. Naturally, as they never had to meet cavalry trained to shock-action only, their leaders made no attempt, in this respect, to bring their men up to the European standard.

Thus I may say that the achievements of our brethren in arms across the Atlantic teach us what may be done by a mounted force that is not much inferior to good infantry, and at the same time has all the mobility of cavalry. Such a force may yet rival the deeds of Sheridan and Stuart in the days to come. But whether that force is to be composed of cavalry alone or of cavalry accompanied by mounted infantry, and whether cavalry can be trained to hold its own on foot without losing something of its dash and daring, are points which, when we take into consideration the deficient training of the mounted forces, the history of the American war does not decide for us. That history, however, shows us one thing, and this is, that if you are going to make great raids on your adversary's communications, to destroy his magazines, and defeat his isolated detachments, or if you intend even to hold his infantry in check with your mounted men alone, your cavalry, when dismounted, must be able to shoot, to manoeuvre, and to attack just as well as infantry.

A sketch of one of these raids will not be out of place here. In March 1865, General Wilson, with some 14,000 cavalry, marched across Alabama and Georgia. He was opposed by General Forrest with 10,000 cavalry, and the important towns were garrisoned by infantry. He marched in thirty days nearly 600 miles, captured three important cities, two of which were protected by very strong entrenchments—which were stormed —and garrisoned, one of them with 7,000 and another with 2,700 men; he crossed six large rivers, fought five battles, destroyed railroads, iron foundries, and factories, and captured 6,000 prisoners and 156 guns. 'In this campaign,' says General Michie, of the Federal army, 'the cavalry, armed with the Spencer (repeating carbine), acted mostly as mounted infantry.'

If your cavalry can be trained to shoot, to manoeuvre, and to attack as infantry, and, at the same time, to manoeuvre well mounted in mass, and if your officers can double the part, or, to paraphrase Mrs. Malaprop, 'become two gentlemen at once'

—the dashing dragoon and the smart light-infantryman—then there is little need for mounted infantry in European warfare.

I think one of the most interesting points connected with the American cavalry is its organisation and working when covering the march or cantonments of the armies, and perhaps a few notes on this subject may induce others, better qualified than I am, to study the manner in which the duties of the cavalry screen were carried out.

At first the Confederates had it all their own way. Not only were they better mounted and more expert horsemen, but the regiments were organised, as early as 1861, in what was practically an independent cavalry division. In the Peninsular campaign of 1862, the Federals possessed the same organisation; but in Pope's campaign, and in the Fredericksburg campaign, in the summer and winter of the same year, the Northern cavalry was attached by brigades to the army corps. In the Maryland campaign, September 1862, the divisional organisation was resorted to, but merely as a temporary measure, and the hastily collected force, like the division in the Peninsula, lacked the cohesion and efficiency which the Confederate division had acquired by long association.

In every one of these campaigns, the superiority of the Southern horsemen, in every branch of tactics, was remarkable. Twice Stuart's division made a complete circuit of the Federal army, and on another occasion rode right into the midst of their cantonments, carrying back as a trophy the commander-in-chief's best uniform.

But in 1863 came a change. General Hooker reverted to the divisional organisation, and his cavalry had several months in which to learn its duties as a single unit under a single hand. From this time forth the mounted arms met on terms of equality; and if the earlier campaigns, like that of 1870, show us not only the value of cavalry well organised and well led, but the helplessness of an army whose cavalry is wanting in cohesion, the later campaigns give us many hints as to the working of the independent divisions. At Chancellorsville Hooker made a fatal mistake. As he moved off his army to attack Lee, he sent nearly the whole of his cavalry to cut his opponent's communications. In the battle which ensued he had with him but one weak brigade; and it is not too much to say, that the surprise and rout of his right wing, and his subsequent retreat, were due to his deficiency in mounted regiments.

Gettysburg, from a cavalry point of view, is perhaps the most interesting campaign of the whole series; and we find here a new feature. The cavalry on both sides was practically divided into two lines, of which the first undertook independent enterprises or endeavoured to bring into action the main body of the hostile horsemen, whilst the second covered the march of the army. This practice obtained during the remainder of the war, and it is remarkable how the two fighting lines continually came into contact, almost every phase of the Gettysburg and the Wilderness campaigns being signalised by some important engagement. In the former the division into two bodies of the Confederate cavalry had the most prejudicial effect. It seems to me that, when the first line under Stuart cut loose from the rest of the army, neither the Confederate staff nor the cavalry leaders in charge of the second line had as yet fully grasped that there is no connection whatever between an independent enterprise and the duty of screening the march; in fact, that it is impracticable to combine the tasks of 'exploration' and 'security.' The Federals worked in very different fashion, and their second line of cavalry, exceedingly well handled, practically decided the issue of the conflict.

One last remark as to the mounted branch of the American armies. From the very outset the Confederate cavalry, untrained as they were, but excellent riders, knowing the country thoroughly, and patriotic and intelligent, did most efficient work upon the outpost line; and I think it is a fair deduction that our own Volunteer cavalry—the Yeomanry—in case of invasion, would, in this respect, prove equally valuable. Nor can I imagine for that force, taking the Confederate troopers as their model, a more honourable and useful rôle than that of mounted riflemen. England is a country which affords even fewer opportunities for purely cavalry combats than Virginia.

It is impossible here to touch on that most interesting of all questions, combined tactics; but I may note one or two points. As with other nations, the Americans seem to have had little difficulty in bringing infantry and artillery into proper adjustment; but, with one single exception, their generals seem to have been unequal to the task of handling the three arms together on the field of battle. The single exception was Sheridan; and his operations, both in the Shenandoah Valley and during 'the last agony' of the Confederacy, are well worth the very closest study.

It is impossible that any soldier should not find the memoirs of such great generals as Lee, Grant, Sherman, Sheridan, Stuart, and many others, most interesting and instructive reading; and in the, unfortunately, rather cumbersome volumes of the 'Battles and Leaders of the Civil War' we have a work which far surpasses any military history that has yet been written. In these books the history of the war may best be studied. There is nothing in them to repel. There is nothing dry. There is romance and sensation enough and to spare; and if we gain nothing else from them, we can at least learn to appreciate the splendid fighting qualities of the American soldier.

THE BATTLE OF GETTYSBURG[1]

(July 1st, 2nd, and 3rd, 1863)

In the first week in May 1863, General Robert Lee, in command of the Confederate Army of Northern Virginia, protecting the approaches to Richmond, the Confederate capital, from the north, had, at Chancellorsville, a few miles south of the river Rappahannock, very decisively defeated a Federal army of invasion more than double his numbers.

This was the third attempt at invasion he had thwarted since the war began, just two years previously; and although his losses at Chancellorsville made it impossible for him to pursue his enemy immediately after the battle, he nevertheless determined, when the Federal army fell back discomfited to Falmouth, beyond the river, to carry the war across the Potomac, the boundary stream between North and South, into the Fed-

---

1. [A lecture to the Aldershot Military Society, February 9, 1893, "The Battle of Gettysburg" was published in *The Science of War*, chap. x. In a lecture before a military audience, Henderson had time to cover only the high spots in the Gettysburg campaign. In a review he later wrote of Longstreet's *From Manassas to Appomattox* (reprinted in the *Southern Historical Society Papers*, XXXIX [April, 1914], 104–17), he analyzed Longstreet's actions on July 2 and 3 in somewhat greater detail. Some of his arguments appear in following footnotes.]

eral territory. Two causes impelled him to an offensive policy. First: Virginia, at no time a rich country, had become almost exhausted by the war, and both the army and the non-combatant population were much straitened for food and supplies. Second: far away in the west, on the river Mississippi, a Federal army was investing Vicksburg, the most important fortress in the South, whose loss would be an irretrievable disaster.[2] He believed it possible that, by threatening Washington, the Federal capital, and the great cities of Pennsylvania, he might induce the Northern President to withdraw the army besieging Vicksburg in order to prevent the Confederates moving with fire and sword through the rich and untouched States of the North. Third: the Federal army, after its recent severe defeat, following on so many successive disasters, was not likely to be strong as regards *moral*, whilst the spirit of the Army of Northern Virginia, after so many victories, was correspondingly high.

His plan of campaign was as follows:—To use the covered approach of the Shenandoah Valley, and its continuation north of the Potomac, the Cumberland Valley, as his line of invasion; to threaten Washington; to defeat the Federal army in a pitched battle; to bring about, if possible, the fall of the Northern capital, and at least to create an important diversion in favour of Vicksburg. Before his operations began, he pointed out Gettysburg as the best point for a battle, as it was so situated that, by holding the passes of South Mountain, he would be able to keep open his line of communication through the Cumberland and Shenandoah Valleys.

⊢ At the beginning of June, his army, consisting of 57,000 infantry, 250 guns, and 9,000 cavalry, in all 70,000 men, was at Fredericksburg, watching the Federal army of 105,000 men, with 300 guns, under General Hooker, at Falmouth, on the opposite bank of the Rappahannock. ⊣

The first step was to remove his troops from Fredericksburg to the Shenandoah Valley; and although the Rappahannock is a broad stream, this movement involved that very dangerous operation, a flank march across the enemy's front.

Now it is by no means sufficient for a student of war to be made aware that a flank march is risky, but what he ought to

2. [Vicksburg was invested on May 18, 1863.]

learn is how to minimise the risk and to escape the danger, for success in war is won by facing danger and not by running away from it. This is one of the great uses of military history. It teaches us, from the experience of the great masters of war, how movements which may be mathematically demonstrated to be vicious, and yet are sometimes absolutely essential to success, may be successfully executed.

The first thing that Lee had to look to was to prevent all information from reaching the enemy. This was provided for by his 9,000 cavalry, who carefully picketed the whole line of the Rappahannock.

Next, he had to remove his troops secretly, and to keep his enemy in ignorance of this movement as long as possible. Now, as the heights at Falmouth looked down upon Fredericksburg this was somewhat difficult. Fortunately the great Virginian forest was very close at hand. Very few of the Confederate camps were visible to the Federal scouts and sentries; the remainder were hidden in the woods.

The third step was to induce General Hooker to march north; thus preventing him, whilst the Confederates were marching into Pennsylvania, from making a dash on Richmond, which was very inadequately defended; and also bringing him out from his strong position and compelling him either to attack the Confederates, or to give them the opportunity of defeating him in detail.

The Confederate army was divided into three army corps, commanded by Generals Longstreet, Ewell, and Hill. On June 2, Ewell's corps, covered from observation by the forest, and screened by the cavalry, marched to Culpeper Court House. Longstreet's corps followed on the 4th.[3] Hill remained at Fredericksburg, in order to induce Hooker to believe that the army was still in position at that point.

It was not till June 10 that Hooker learned what was going on. He immediately extended his line along the Rappahannock, his right resting at Bealeton, north of Culpeper. Hill was still at Fredericksburg. On the 9th the Federal cavalry, three divi-

3. [Longstreet's corps left Fredericksburg on June 3; Ewell's corps followed during the next two days and reached Culpeper on June 7 (*Official Records*, Ser. I, XXVII, Part II, 293, 439).]

sions, had driven in the Confederate pickets, crossed the Rappahannock, and encountered Stuart's cavalry, at Brandy Station.[4] An indecisive engagement resulted. But Hooker discovered that a large part of the Confederate army was at Culpeper, and determined to reinforce his right. On the same day, Ewell was sent into the Shenandoah Valley, to capture Winchester, and to create the impression that a flank movement against Washington, an operation which Lee had made most effective use of on three previous occasions, was in contemplation.

On June 12, Hill was still at Fredericksburg with his 20,000. Within reach, on the opposite side of the Rappahannock, were no less than five Federal *corps d'armée*, numbering 70,000 men. Further to the right, opposite Chancellorsville, was another Federal corps of 15,000 men; and still further, yet another, round Bealeton, facing Longstreet's 20,000 at Culpeper. Ewell had reached Front Royal. Thus the three Confederate corps were each forty miles apart, and opposite the space between Hill and Longstreet were massed, on a front of some thirty-five miles, 100,000 Federals. According to all the rules of war, Hooker ought to have been easily able to deal with Hill and Longstreet in detail, for a march of fifteen miles, at furthest, would have placed him between them in overwhelming force. He had quite enough information to make it clear what an excellent opportunity the apparent rashness of the Confederates had given him, and he sent back to Washington for permission to cross the Rappahannock, defeat Hill, and move rapidly on Richmond.

He was refused; and ordered instead to defend the approaches to Washington.

And Lee knew that he would be refused, and this was the secret of the seemingly foolhardy position in which the Confederate army was distributed in face of superior numbers.[5]

4. [For account of the battle of Brandy Station, see chap. V, pp. 214 ff.]

5. [Henderson here assumes greater knowledge of Lee's motives than do subsequent biographers of the Confederate commander. Freeman states only that Lee believed his movements would strengthen the hand of those in the North desiring a compromise peace (Douglas Southall Freeman, *R. E. Lee* [New York, 1935], III, 33). Maurice advances sound military reasons why Lee's dispersed order of march involved reasonably few risks (Sir Frederick Maurice, *Robert E. Lee the Soldier* [Boston and New York, 1925], pp. 198–99).]

How had he come to be possessed of this information? It was not through his cavalry patrols, not through prisoners, not through his spies; but through his knowledge of the character of the Commander-in-Chief of the Northern armies. He knew well what apparent risk he might run with absolute impunity. He knew that the superior numbers of his adversary, and his own dangerous position, were factors in the problem of but small account. He knew that in war moral means, according to Napoleon, are three times more effective than physical means, that is, than numbers, armament, and position; and it was on the former that he now relied.

War is more of a struggle between two human intelligences than between two masses of armed men; and the great general does not give his first attention to numbers, to armament, or to position. He looks beyond these, beyond his own troops, and across the enemy's lines, without stopping to estimate their strength or to examine the ground, until he comes to the quarters occupied by the enemy's leader; and then he puts himself in that leader's place, and with that officer's eyes and mind he looks at the situation; he realises his weakness, tactical, strategical, and political; he detects the points for the security of which he is most apprehensive; he considers what his action will be if he is attacked here or threatened there, and he thus learns for himself, looking at things from his enemy's point of view, whether or no apparent risks are not absolutely safe.

This is what Lee had done before he ventured on distributing his army corps along so wide a front. He looked beyond his own army, beyond the enemy's camps, beyond the tent of their commander—the man who was eager to profit by the opportunity he offered him—and across the great river which divides Virginia from the North. Over the river he saw Washington and the President's house, and in the President's chair sat a man called Abraham Lincoln, by virtue of his office, civilian though he was, Commander-in-Chief of the Federal armies, and the motive power of the forces which Hooker commanded in Virginia. It was this motive power that Lee attacked. It was against this man that he fought, and not against the masses on the Rappahannock. He knew well that political necessities were Lincoln's chief preoccupation. He knew his apprehensions for the safety of the Union capital. He knew that a threat against Washington was an infallible specific—he had tried it already —for making the enemy divide his enormous forces, detach whole army corps for service round the city, and for compelling his armies to withdraw from Virginia, whether they were badly beaten or not. So, when he sent Ewell to the Shenandoah Valley, an advance from which, as is evident from the map, would threaten the communications of Washington with the more northern States, he was morally certain that Lincoln, the motive power of Hooker's army, would draw that army back to protect Washington instead of pushing it forward against Hill.

In exact accordance with this anticipation, Hooker fell back on the night of the 13th, and changing front to the right, occupied Leesburg, and the passes of the Bull Run mountains.

The whole Confederate army now crossed the Blue Ridge, Stuart, with the cavalry, remaining in rear to watch the enemy and to block the passes. On the 23rd began the passage of the Potomac.

Hooker, at Leesburg, covered the fords of the Potomac north of that town, and threatened the flank of the Confederate advance. On the 25th he heard that Lee had crossed the river near Leetown,[6] and immediately followed suit, intending to operate against the enemy's rear. The President, however, objected to the plan of campaign, and the general asked to be relieved of his command. He was succeeded by Meade, like Lee, an officer of the United States corps of Engineers, who took over his new duties on June 28, and moved the army immediately on Frederick. Lee's advanced guard had by this time reached Carlisle and York, and was threatening Harrisburg. Meade, moving rapidly northward, resolved to force Lee to battle before he could cross the Susquehanna river.

We have now to deal with a certain resolution of General Lee's which had a very startling effect on the campaign.

On June 23, on the day on which the passage of the Potomac began, General Lee gave his cavalry commander, Stuart, who up till that time had been guarding the Blue Ridge gaps against the Federal cavalry, permission to move round the rear of the

6. [Longstreet's corps crossed the Potomac at Williamsport and Hill's corps at Shepherdstown. Ewell's corps by this time was somewhere in the vicinity of Chambersburg (*Official Records*, Ser. I, XXVII, Part II, 307, 442–43).]

Federal army, then at and about Leesburg, to cross the river, and doing what damage he could to join the advanced guard of the Confederates near the Susquehanna. He was to employ three brigades, leaving two brigades behind, which were to watch the Blue Ridge passes until the infantry was on Northern soil, and then to join the army. Of the two remaining cavalry brigades, one was with the advanced guard, the other well away on the left flank, on the far side of the Cumberland Valley.[7]

Now Stuart had been in the habit, in former campaigns, of taking his division for a trip round the enemy's army, cutting their communications and acquiring information. It does not appear that great good invariably resulted from these enterprises. The American railways, if easily destroyed, were just as easily repaired, and merely riding across the enemy's communications is a very different thing to placing an army astride or on the flank of his communications. The latter course almost invariably compels him to turn back on the intruder; the former inflicts but temporary discomfort. Still Stuart had always been successful in these raids; by his extraordinary energy, activity, and tactical skill, he had won Lee's confidence, and his superior seems to have acquiesced without question in the suggestion that, with the larger half of his command, his trusted cavalry leader should separate himself from the rest of the army at a critical time.

That the cavalry did do a certain amount of damage in this raid is true; but it may be doubted whether they delayed the northward march of the Federal army for a single hour; and, owing to the fact that Hooker crossed the Potomac sooner than either Stuart or Lee expected, instead of crossing the river two days in advance of the Federals, they did so two days behind them, and did not join the advanced guard until July 3, with both men and horses much exhausted.

Meanwhile the two brigades of horsemen left behind proved

7. [Lee did have four brigades of cavalry at his disposal, two left behind by Stuart and two recently arrived from West Virginia. The total strength of these four brigades was in excess of 7,000, and they were located as follows: the two brigades from Stuart's division were assigned the task of guarding the mountain passes on the right flank and rear of the Confederates, a third brigade operated on the left of the army, and the fourth accompanied Ewell's corps in Pennsylvania. But Stuart had taken the cream of the cavalry with him, and Lee had no cavalry available to perform the essential duties of reconnaissance.]

insufficient to keep a watch on Hooker and to break through his cavalry screen.

Stuart marched on the 24th. On the night of the 25th Hooker began to move from Leesburg. But it was not till the night of the 28th that Lee was made aware, by a spy, that the Federals had crossed the Potomac. Believing that their army was still south of the river, he had allowed his army corps to move in very open order.

On the 28th, Ewell's corps reached Carlisle and York; the other two were near Chambersburg, from thirty to fifty miles in rear.

On the night of the 28th, hearing the Federal advance, Lee immediately called up the two brigades of cavalry from the Shenandoah Valley, about fifty miles distant, and ordered his army to concentrate at Cashtown, nine miles west of Gettysburg.

The important circumstance to notice is, that from the time the Confederate infantry crossed the Potomac until after the battle of Gettysburg had been fought and lost, Lee had not a single cavalry soldier between himself and the enemy. For nearly four days he remained in ignorance of the Federal movements; he did not know that their army had crossed the river, and he had consequently allowed his three corps to separate so far that it took four days to effect their concentration.

On the 30th, Hill reached Cashtown, and the rest of the army was not more than fifteen miles distant. The leader of his advanced guard sent a brigade on to Gettysburg to procure a supply of boots, and this brigade returned with the information that the town was occupied by the enemy.

In nearly every book on tactics we have instances of the great use of cavalry in screening the front and reconnoitring. At Gettysburg we have an instance of this screen being altogether absent; and I think the difficulties of the general, arising from this absence, will illustrate how completely the other arms are paralysed without the aid of the cavalry.

That very afternoon a Federal cavalry division under General Buford, scouting far ahead of the army, had entered Gettysburg. This division, all told, did not exceed 4,000 men, and the nearest infantry support was over fifteen miles distant.

Now Gettysburg was important in two ways. It was tactically a strong position, commanding the approaches from the west and north, and it was strategically most important, for it was the nucleus of several good roads, leading to the Susquehanna and the Potomac, to Philadelphia, Baltimore, Frederick, and Washington. The Federals, then, were most fortunate in anticipating the Southerners in the occupation of this point of vantage.

I use advisedly the term 'fortunate,' for on the morning of the 30th, the greater part of two Confederate army corps were within nine miles of Gettysburg, Hill at Cashtown, and Ewell, returning from York, at Heidlersburg, on the opposite side of the town; and by making a little haste, either or both of these corps would have been firmly established on the heights of Gettysburg before the Federal advanced guard arrived. Had the Confederate cavalry been present, scouring the country to the front, the enemy's approach would have been reported, and measures might have been taken to anticipate him in securing this important point.

This, then, was the first untoward circumstance which arose from the absence of all reconnaissance. It was followed by others.

On the morning on the same day, Hill, as I have said, had sent a brigade, probably not less than 2,000 strong, to get supplies at Gettysburg. The Brigadier, as he neared the town, saw a Federal force advancing to meet him. This force consisted only of cavalry, Buford's division, and as the strongest of its three brigades had been detached to Mechanicstown, it did not number more than 2,000 men and a couple of horse artillery batteries.

Had the advance of the Confederate brigade been covered by cavalry, in all probability the strength and composition of the enemy's force, and also whether it was supported, would have been ascertained; and the Brigadier would have been free to contest with the cavalry the possession of the Gettysburg position, or at least to have sent back reliable information.

What followed? The Confederate brigade withdrew to Cashtown, reporting the advance of a large hostile force on Gettysburg; and next morning, July 1, General Hill went forward with two divisions of infantry to ascertain the strength of the enemy.

These two divisions found the Federal cavalry dismounted, holding a strong position in front of Gettysburg, and gradually drove them back upon the town. Meanwhile, between 10 A.M. and 1.30 P.M. two Federal army corps arrived, and the Confederates were in their turn pushed back. Then at 2.30 P.M. up came Ewell from Heidlersburg, and a general advance drove the Federals through Gettysburg at 4 o'clock with very heavy loss.

Near the close of the action General Lee arrived upon the field, and the whole of his army was rapidly closing up. But it was still far from being fully concentrated, and so exhausted were the troops in immediate contact with the enemy, so strong the position to which the Federals had retired, a mile south of the town, and so uncertain the estimate of their numbers, that the Confederate general made no effort to follow up his success. He directed the necessary preparations to be made for an attack the next morning as early as practicable.

Thus ended the first day's battle, in which about 22,000 men on each side were engaged, and which resulted in a Confederate victory.

There are several points which may here be noticed; the first regarding the Confederate strategy on this day.

To begin with. When Lee started upon the campaign, he had not intended to deliver an offensive battle at so great a distance from his base of operations, but, owing to the absence of his cavalry, and the engagement brought about by Hill's reconnaissance, he had now no other course open but to attack the enemy as vigorously as possible. It would certainly have been more promising of success had his inferior army of 70,000 men been able to await, in a carefully prepared position, the attack of the 100,000 Federals led by Meade.

Certain critics of the campaign, amongst them the Comte de Paris, the historian of the war, and General Longstreet, commanding one of the Confederate army corps, hold very different views. They assert that Lee had three other courses open to him, each of them more promising than the one he actually adopted.

1st, to retire to the passes of the South Mountain, and thus to compel Meade to attack him in a very favourable position.

2nd, to await attack in his present position.

3rd, to move round the left flank of the Federal position and to interpose between the Federal army and Washington, taking up a strong position; and if Meade refused to attack, to move back in the direction of Washington, which threat to the capital would probably induce the Northern general to do so.

In his report of the battle the Confederate Commander-in-Chief disposes of the first two of these suggestions very summarily. His army was living on the country, and it would have been exceedingly difficult to subsist 70,000 men, occupying a stationary camp, in face of a numerically superior enemy. No district, however rich, can supply a large army for more than forty-eight hours, and the greater part of the army had just passed through the district east and west of Cashtown by easy marches.

As to the last proposal, which was strenuously urged after the first day's battle by Longstreet, Lee, according to one of his chief staff officers, pronounced it, under the circumstances, impracticable.

Now, what were the circumstances that thus paralysed his army and his own great skill in daring manoeuvres? Why was a flank march, possible in front of Hooker in June, impossible in front of Meade in July?

⊦The answer is simple—the absence of the cavalry.⊣

One of the chief requisites for a flank march is that it should be made with the greatest rapidity. What speed was possible if the infantry divisions were compelled to reconnoitre themselves to front, flank, and rear, halting at every alarm, harassed by the hostile horsemen? How was Lee to ascertain whether the enemy had not a force posted to his left rear, ready to crush the head of the turning column?

We have only to turn to the disastrous march of MacMahon's army, culminating in the terrible defeat of Sedan, to understand the difficulties and danger of a flank march without cavalry to screen the movement.

I need hardly say that the other alternative, a retreat through the South Mountain, was never entertained for a moment. To withdraw by narrow roads in face of superior numbers would have been no easy matter. Moreover, a retreat would have left to the enemy all the moral results of victory, and would have been everywhere interpreted, by foreign nations as well as by the Northerners, as a confession of weakness on the part of the Confederate leader and of the Confederate Government.

Lastly, it is evident that had Lee's army been closely concentrated, which it would have been had he received early information of Hooker's march northward; he would have been able to seize Gettysburg and to inflict an annihilating defeat on the two corps which formed the Federal advanced guard.

As to the Federals. We may first of all notice the brilliant initiative of General Buford, the cavalry commander, who, on reaching Gettysburg, and recognising the importance of the position, determined to hold it, although hostile infantry was visible, until his own infantry came up. Second, the value of cavalry who were so trained as to be able, when dismounted, to hold in check a superior force of infantry for two hours, and to give time for the arrival of reinforcements. I may notice that this same cavalry, later in the day, when the Federal line was giving way, was ordered to charge the victorious enemy pressing forward in pursuit. The charge was never made—probably the nature of the ground, the numerous woods, walls, and fences, forbade it—but the division formed up with every intention of charging, and it is said that the Confederate battalions formed square, and so lost much previous time. Third, the judicious selection of a position by this same officer, not on the crest of the ridge immediately south of Gettysburg, but along the banks of Willoughby Run, more than a mile west. He recognised that the ridge to the south was the true position; and that as he would certainly be sooner or later forced back, it would be better to leave it to be strongly occupied by the remainder of the army. As it turned out, when the troops west of Gettysburg were forced back on the morning of July 1, they found the ridge occupied and entrenched. As we have seen, General Lee judged it too formidable to attack the same evening.

There is another Federal officer whose conduct calls for the highest commendation. This was General Hancock, commanding the 2nd corps. To appreciate his action we must turn to what General Meade had been doing since he started on his northward march from Frederick on the 29th. Till the evening of July 1, the first day of the battle, he was ignorant of Lee's whereabouts. All he knew was that the Confederate army was somewhere between Chambersburg and Carlisle, and that it was now moving southwards. His own army corps were dispersed over a wide extent of country east and south of Gettysburg. But he knew enough of Lee's movements, and whilst Hill and Ewell were converging on Gettysburg for the assault on his advanced guard, he was issuing orders for his chiefs of engineers and artillery to select a field of battle, covering Lee's lines of approach, whether by Harrisburg or Gettysburg, indicating the general line of Pipe Creek as a suitable locality.

But on receiving news of the fight at Gettysburg, he sent Hancock to the battlefield, directing him either to bring the two advanced corps to Pipe Creek or to prepare for a general engagement at Gettysburg.

As soon as the action with Hill and Ewell was over, and the defeated Federals were firmly established on the ridge south of the town, Hancock sent back to Meade, whose headquarters were thirteen miles in rear, informing him that the position was a very strong one. Moreover, he kept his men behind their entrenchments, without taking any step towards retiring to the line of Pipe Creek. This, as it turned out, was a most momentous decision, and I think that the courage of the general who, in command of a defeated force, confronted by superior numbers, and aware that the supporting army corps were much scattered, refused to abandon the strong and formidable position he occupied and to leave to the enemy the moral results of a victory culminating in the retreat of the vanquished, is well worth notice.

Meade, relying on Hancock's soldierly instinct, and appreciating his motives, hurried the whole of his corps, scattered as they were, to the front, and at midnight rode forward to the field.

By forced night marches his troops pushed on, but at daylight next morning only four of the seven corps were present, and two of these had been very roughly handled on the previous day. By eight o'clock two more had come up, making in all some 65,000 men.

At daylight, however, there were no more than 40,000 present, and it is very evident that the Confederate attack should have been made at that hour. It is also evident that the Federal corps, like the Confederate army, had become separated by too wide intervals in their advance; and, in the absence of information, concentration should be an invariable rule.

During the night, Lee had learned from prisoners that only a portion of the Federal army occupied the opposite ridge. It was clear that an opportunity presented itself of dealing with the enemy in detail; and the meanest capacity must have grasped the advantage of storming the strong position south of Gettysburg before it should be occupied in overwhelming strength.

⊦ Now Lee's own orders to his lieutenants had been to attack 'as early as practicable.' But as a matter of fact the attack was not made until 4 P.M., just eleven hours too late. ⊣

On this circumstance, which has given rise to much unpleasant controversy amongst the surviving generals of Lee's army, I shall make no comment beyond saying that it was unfortunate that the attack should have waited on the movements of Longstreet, the general who had so strenuously advocated the flank movement to turn the Federal left.[8]

⊦ Moreover, there was very indifferent Staff work done on this morning in Longstreet's corps, one of his divisions taking a wrong road, and much delay being caused by the fact that the roads were not reconnoitred previous to the march.⊣

As to the fighting on this day little need be said. The Federals were strongly posted from Cemetery and Culp Hills, on the right, to a point west of the ridge, on which stands the Peach Orchard, on the left. The orchard, standing on a rise a good deal lower than Cemetery Ridge, had been occupied, not on General Meade's authority, for he intended his left to rest on the Round Tops, but on the initiative of the general commanding the left wing, and, as may be seen from the map [Map VI], it was salient to the rest of the line, and much nearer to the Confederate front than the right flank. Lee's plan of attack was as

8. ["He was aware that Lee was anxious to attack as early as practicable . . . [and] that an early attack was essential to success. He was aware how the commander-in-chief desired his divisions should be placed; and yet until he received a definite order to advance did absolutely nothing. He made no attempt to reconnoitre his line of march, to bring his troops into position, or to initiate the attack in accordance with the expressed intentions of his superior" (Henderson, "Review of General Longstreet's Book," p. 113).]

MAP VI
The Campaign of Gettysburg

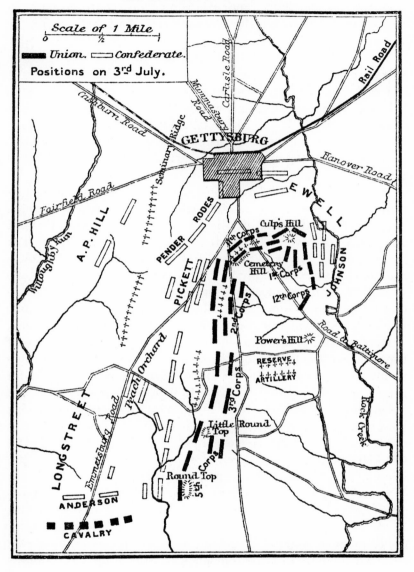

follows: Ewell, from the north, that is, from Gettysburg, and the height to the east, was to attack Cemetery Hill. Longstreet on the right was to attack the Peach Orchard position, turn the Federal flank, and, wheeling half-left, to advance in the same direction up the Emmetsburg road, rolling up the Federal line from left to right.

The two attacks were to be made at the same moment, and this part of the programme was carried out.

Ewell assaulted the Federal right in two columns. That on the left, Johnson's division, which moved on Culp's Hill, was fairly successful. When night fell, Johnson's troops had possession of a line of Federal entrenchments, and held on to this position during the night. But the attack on Cemetery Hill was a failure.

It was made by two divisions, one from the east of Gettysburg, to be supported by another which had to advance through the town itself. The first division, under General Early, had but 700 yards to traverse before it reached the Federal lines. The second, under General Rodes, had to move out of the town, then to deploy, and finally to move over a space of nearly 1,400 yards. The consequence was that Early attacked and was successful, but the co-operating column failed to come up in time to enable him to meet a counter-attack, and he was driven back.

Here, again, it is impossible not to criticise the working of the staff. On the field of battle, to see that the combined movements of the large units are made with due consideration for time and space is the most important duty of the staff.

On the Confederate right, Longstreet succeeded in driving back the Federals from the Peach Orchard line. But he was unsuccessful in rolling up their line towards Cemetery Hill. The Confederate right was already in position to attack Little Round Top, the key of the position, when a Federal general, Warren, Meade's Chief of Engineers, reached the hill with orders from Meade to examine the condition of affairs. From this height he saw, in the long line of woods west of Emmetsburg road, the glistening of gun-barrels and bayonets, and, promptly realising the situation, he sent back to Meade for a division at least. The situation, he says in his report, 'was almost appalling.' Fortunately, before the Confederates could reach this hill, where they would have been established in rear of the Peach Orchard, and whence they would have enfiladed a great part of the ridge, a Federal brigade, with some batteries on their way to reinforce the line in front, came up and were ordered by Warren, on his own responsibility, to occupy Little Round Top. Here a heavy struggle took place, and although Longstreet drove back the enemy from the Peach Orchard, he was unable to turn their flank, for reinforcements were rapidly brought up to the rocky ridge and ravine called the Devil's Den, and the Federals made good their withdrawal to Cemetery Ridge.

I may add that the Federal infantry, bravely as it fought, seems to have owed its safety to the devotion of the gunners, who showed most remarkable gallantry in covering their retreat. One battery lost all its officers but one, six out of seven sergeants, twenty-eight men out of one hundred, and sixty-six horses out of eighty-eight. Moreover, a line of five-and-twenty guns, hastily moved up to the ridge in rear, although unsupported by infantry, did much towards checking the Confederate pursuit.

There is one point connected with this attack which calls for particular comment.

I have already stated that Round Top was the key of the position; and it is evident that had the Confederates once occupied this commanding height, the Federal troops, when forced back from the Peach Orchard, would have been compelled to retreat towards Cemetery Hill.

When Longstreet's line got into position, his right brigade was well in front of the Emmetsburg road, at an oblique angle to it, and this brigade was supported by a second, 200 yards in rear. As soon as the troops took up their place, the commander of the advanced brigade, General Law, sent off a patrol of six men to ascend the steep and densely wooded slopes of Round Top, and to locate the extreme left of the Federal line.

Before the attack began, one of the men came back at the double, reporting that Round Top was unoccupied, and that there were no Federal troops in rear of the hill. This intelligence was corroborated by some prisoners who were just then captured. The Brigadier immediately rode over to his divisional

The Battle of Gettysburg 241

commander, and pointed out the ease with which the Federal left might be turned. The divisional commander coincided fully with his views, but declared that his orders were positive to attack in front. On the Brigadier protesting, the divisional commander sent an aide-de-camp to General Longstreet. An order was sent back which was interpreted to mean that the original plan of attack was to be followed out to the letter. The right brigade, therefore, moved forward against the Devil's Den, cleared that, and when it afterward moved against Round Top, found it occupied and was beaten back.

We can only say that it seems unfortunate that the question whether the attack on Round Top was advisable or not should have been submitted to the general that had so strongly advised Lee not to attack the Gettysburg position at all.

His summary message to the divisional commander to carry out the original plan at least lays him open to the suspicion that, although he was prepared to obey orders, it was like a machine and not like an intelligent being. There was no question of acting on his own initiative even, and of taking it on himself to modify his instructions. The Commander-in-Chief was close at hand, and he might have communicated with him at once, just as his subordinates had done with himself.[9]

9. ["We do not for a moment believe that General Longstreet can fairly be charged with deliberate disloyalty to his superior. He set out on the campaign with a false idea of their relative positions. . . . He had no mandate from the government to act as Lee's adviser. He was merely the commander of an army corps—a subordinate, pure and simple; and yet he appears to have entered on the campaign with the idea that the commander-in-chief was bound to engage the enemy with the tactics that he, General Longstreet, had suggested. . . . When the enemy was encountered, his irritation at the rejection of his advice was such that he forgot his duty. . . . The question is not whether the maneuvers suggested by Longstreet would have been more successful than those executed by General Lee, but whether . . . [Longstreet] did everything which lay within his power to carry out, loyally and unhesitatingly, the wishes and instructions of the commander-in-chief of the Confederate army. . . . If he moved only under compulsion, if he deliberately forbore to use his best efforts to carry out Lee's design, and to compel him to adopt his own . . . (that he did so seems perfectly clear) . . . it is impossible for any sane soldier to justify such conduct. . . . His error was amply atoned at a later period; and had he frankly confessed that his temper got the better of him on July 2 and 3, we might easily overlook the one blot on the career of a gallant soldier. But his endeavors to clear his own reputation by assailing those of others, together with the bitterness of his recriminations, serve only to alienate sympathy and destroy respect. General Longstreet did splendid service for the South. He has been subject to the merciless attacks of many enemies. He has been assailed with accusations which are utterly without founda-

On this same evening of the battle of July 2, there was a very curious exercise of initiative, a very marked assumption of responsibility, on the part of two Federal officers. One of these was General Warren, who, on seeing Round Top without a single bayonet on it, dashed down the hill and ordered up the first regiment he came across. The other was the regimental commander, who, although following the leading battalion of his brigade, on receiving an urgent demand for assistance from a senior officer of the general staff, accompanied by a brief explanation of the situation, broke the line of march of his brigade without hesitation, and marched straight up the hill, arriving in time to secure its possession to the Federals.

In Germany, where the advantages of the initiative are most highly appreciated, this question of how far a commander, coming up in support with orders to move to a certain locality, is justified in answering urgent appeals for assistance from another locality altogether, and in departing from his original orders, is often very warmly discussed.

An incident occurred at the battle of Woerth, in 1870, which has been made the text of a long discussion in a German study of that battle; a study which is well worth reading, and which, for the consolation of those who do not read German, was admirably paraphrased, by Colonel Lonsdale Hale, in the 'Contemporary Review' for June 1892.

In this case it does not appear that the need of support was absolutely necessary in order to save the day. There was nothing in the situation which clearly indicated, as at Round Top, that if the supporting troops obeyed their original orders the battle would be lost.

The third day of Gettysburg dawned on two armies that still stood face to face on equal terms. The Confederates had carried the Peach Orchard, and Johnson's division was established on Culp's Hill, but the Federals occupied a stronger position than on the previous day, the line from Cemetery Hill to Round Top. Their strongest corps, which had not come up until the

---

tion; and it may seem harsh in the extreme to criticize the veteran's defense of his military conduct. But where historic truth and great reputation are at stake it is impossible to be silent" (*ibid.*, pp. 110–17, *passim*).]

evening of the 2nd, had not yet been engaged; and their troops were concentrated in a horseshoe which did not measure more than two and a half miles. They had indeed suffered a severe repulse the previous afternoon. But the Generals, assembled at a council of war after the battle had ceased, had resolved, with scarcely one dissentient voice, to maintain their ground despite their heavy losses; and the morning of one of the most momentous days in American history saw their volunteer soldiers, worn and exhausted as they were with two days' fighting, which had been all against them, outflanked on one wing, and with an enemy before them who had beaten them—or rather their generals—in battle after battle, still resolute, confident, and even cheerful. By all the rules of war they should have been demoralised and unnerved. Yet they were never in better spirit for the fight than on this third day of battle, with their line of retreat seriously threatened by the presence of Johnson's division in rear of their right wing and with nothing but disaster during the past two days to look back upon. Surely they had inherited the best quality of British soldiers. They refused to acknowledge that they were beaten.

The Confederates, flushed with the partial triumph of the preceding day, had no helpless prey before them. When the light broke on the Cemetery Ridge, showing the Northern batteries and battalions still in position, covered by breastworks and stone walls, and commanding the long open slopes to the westward, it was evident that the hardest part of the task was yet to be accomplished. And to make matters worse, the army was badly placed for attack. From Johnson's left on Culp's Hill to Longstreet's right below Round Top, the front covered no less than five miles, more than twice the front occupied by their opponents, who were also superior in numbers.

Nor was it possible to shift troops from one flank to the other. The roads by which they would have to pass were not only visible from Cemetery Ridge, but were commanded by the Federal artillery. The army, owing to the absence of the cavalry, had blundered into battle on the first day. Ewell had then attacked from the north, and it was almost impracticable afterwards to contract the line to a reasonable length. Such an extent of front, manned by only 60,000 men, swallowed up almost all reserves;

and on the morning of July 3, Lee had one of the hardest problems to deal with that was ever proposed upon the field of battle: Which part of that long extended line should be thrust forward to make the decisive stroke, which was to annihilate the last army of the Federals in the east, and drive the Northern Government from the capital?

So confident was he in the powers of the gallant men he had led so often to victory that, difficult as was his task, Lee never seems for a single moment to have despaired of success. Yet the day opened ominously. As the sun rose, a vigorous attack of the Federals on Culp's Hill, prepared during the night, drove Johnson's division in panic down the hill. But the great Confederate general was not disconcerted by the mishap. It would have been scarcely possible to support Johnson with sufficient force to make an attack on Culp's Hill decisive, and his mind was already seeking to find a point where he could attack with all his strength, and where, to the Federals, defeat would mean annihilation. The right flank of the enemy was secure, for he could not move troops in that open country to attack it, and it was far from their line of retreat. The left flank rested on the impregnable position of Round Top, and he dared not weaken his line to turn it. There remained only the centre, and he determined to try Napoleon's decisive stroke.

The action began at 1 P.M., by which time the Confederates had brought 140 guns into line from opposite Gettysburg to the Peach Orchard ridges. Their fire was answered by ninety Federal guns upon the opposing crest. At 3 o'clock the hostile artillery ceased fire. Eleven ammunition wagons had been blown up, but the losses had not been heavy; in fact, the fire was more dangerous behind the ridge than on its crest. The fire was not concentrated, but scattered over the whole field. The Federal chief of artillery, however, found his ammunition was running low, and resolved to keep his remaining rounds for the assault which he knew must follow. The Confederates, on finding that the enemy had ceased fire, immediately moved forward to attack, thus making that too common mistake of neglecting to bombard the enemy's infantry when his guns have been silenced. During the artillery duel the Federal infantry had been lying behind the entrenchments and stone walls. They had suffered but little loss; they were in no wise demoralised, and were perfectly ready to defend the position to the last.

Lee's scheme of attack was this. Longstreet, who had been reinforced in the night by a fresh division of 5,000 men, Virginians, some of the best troops in the army, and led by General Pickett, one of the most daring amongst the Confederate officers, was to send in three divisions, one of his own and two of Hill's, numbering about 15,000 bayonets, and the flanks of the column were to be protected by the advance of the artillery. Nor was this all, at least according to the testimony of Lee's staff officers. He intended, according to them, that the attack should be supported by Longstreet's two remaining divisions, and the general was authorised to employ another of Hill's divisions if necessary, in all 30,000 men. This General Longstreet denies, but it is remarkable that the two divisions of his right wing, posted opposite the Federal left, never moved a step forward nor were ordered to make any attempt whatsoever even to demonstrate in favour of the attacking column.

The attack, then, was made by 15,000 men in two lines, Pickett leading, Pettigrew in short echelon (100 paces) to the left rear, and Wilcox's brigade to the right. The distance the men had to traverse was nearly 1,200 yards in width. The ground was open, and intervening between them and the enemy were several fences, a field of corn, a tiny brook, and then the open slopes to the Federal position, covered on the crest by earthworks and stone walls.

Notwithstanding the strength of the position they were to storm, and the terrible fire at that range which the Federal artillery, coming into action again as they advanced, poured into their ranks, Pickett's Virginians advanced with a steadiness and precision which called forth the generous admiration of their gallant enemy. Only the skirmishers in front used their rifles, and the long lines in rear pressed forward without a check. Thrown somewhat into disorder in clearing the fences of the Emmetsburg road, they wheeled half-left at the house which stood in their path, and moved straight up the slopes in the direction of a conspicuous clump of trees. The long lines of Federal infantry opened on them in front. The guns, loaded with canister, tore great gaps through the crowded ranks, and from the slopes of Little Round Top they were enfiladed by more than one battery. As they approached the ridge their lines were torn by incessant volleys of musketry, and the second line crowded in upon the first. Under the heavy fire the supporting division on the left had given way, and a Federal brigadier, throwing forward a regiment with ready judgment, enfiladed Pickett's line. Yet with unfaltering courage the Virginians broke into the double, and with an irresistible charge went through and over the stone walls which confronted them, driving back their defenders, from flank to flank, and planting their colours on the summit of the ridge.

But they were few in number; and, as in the history of too many famous charges, at the moment of their success they looked back vainly for support. Not a single Confederate bayonet, save in the hands of wounded or retreating men, was between them and the ridge from which they had advanced, 1,200 yards in rear. Fiercely they struggled to maintain their position, but their courage had been thrown away. The Federals, though driven back, had not lost heart. The defence was as stubborn as the attack was dashing. Fresh regiments came thronging up, and within ten minutes Pickett, with the relics of his brave 5,000, was retreating down the slope. It may be a fitting climax, that magnificent charge, to a battle never surpassed for desperate fighting; and it seems according to the fitness of things that the two commanders should have tacitly agreed to bring the conflict to a close. Meade made no attempt to initiate a counter-attack; and during the night, slowly followed by his adversary, Lee fell back through the South Mountain passes, and across the Potomac into Virginia.

The losses in the battle amounted to over 20,000 on either side, and it is said that Pickett alone lost six-sevenths of his strength.

There are two points to be noticed in connection with the third day's battle. First, the want of co-operation. What sight more curious than to see two armies, each of over 60,000 men, watching in breathless silence the advance of 15,000? Why were not Ewell's troops attacking on the left and Longstreet's remaining divisions on the right? We can only say that some one blundered. Again, remember that Pickett's flanks were to have been protected by the advance of the artillery, but the Confederate batteries, when the artillery duel ceased, had expended nearly all their ammunition, and this all important circumstance was never reported to General Lee.

I have said very little of the tactical use made by General Meade of his formidable position, but I would commend to anyone who may at some future time care to study this battle in detail, to notice particularly how skilfully he used his reserves, transferring them from point to point and throwing them without hesitation into the fight at the point where they were most needed, and how he was assisted in so doing by the small front and great depth of his position.

There are still a few points on which I should like to touch.

As regards the employment of the cavalry in the battle of Gettysburg, there are one or two incidents worth notice. On the third day the Federal cavalry south of Round Top did good service, both dismounted and mounted. Dismounting and occupying some stone walls they compelled Longstreet to detach a force to his right in order to hold them in check; and, mounted, they made a gallant charge across very difficult country soon after Pickett's charge had been repulsed. This charge was certainly attended by heavy losses. But it threw the Confederate infantry on this wing into confusion, and had it been followed up by the Federal infantry on Round Top [it] might have had a startling effect.[10] The cavalry, however, was unsupported; but the confusion it created in the Confederate ranks, difficult as the ground was over which it charged—rocks, timber, and stone walls—leads up to the reflection that had Pickett been supported by cavalry the counter-attacks on his flank and the rallying of the Federal regiments when he carried the ridge would, at least, have been much interfered with. But Lee had no cavalry available. Stuart was well away on the left wing, north-east of Gettysburg, engaged with the main body of the Federal horse. He made a vigorous charge about the same time that Pickett moved out, evidently with the design of spreading panic in rear of Meade's army and so aiding the frontal attack, but was beaten back in a hand-to-hand fight.

10. [This was the charge of two Union cavalry regiments under the command of Brigadier General Elon J. Farnsworth.]

In the wars of the future, when two armies are drawing near each other, the independent cavalry divisions will come into contact, and they will concentrate for a cavalry battle, possibly leaving either the front or flanks of their infantry uncovered, and affording an opportunity for the enemy's army to approach unobserved. This possibility is well worth notice, for at the last French manoeuvres at which I was present, an incident occurred which showed that when the cavalry division is well out in front the commanders in rear feel a sense of security which is not always justified, and that they are prone to think themselves relieved of the necessity of reconnoitring their own line of march. The incident I refer to was the complete surprise of an entire infantry division by a brigade of cavalry and a horse artillery battery, owing to the absence of the very small force of divisional cavalry, a squadron only, in another direction, and the belief that the independent cavalry were watching the flank. As a matter of fact they were on this flank, but very far to the front, and whilst they were heavily engaged with infantry the enemy's brigade had worked round to their rear, and appeared on the flank of the advancing column. In the Gettysburg campaign, I cannot help thinking that Stuart forgot for once that to cover the march of the army and to send in timely information are services of far greater importance than cutting the enemy's communications and harassing his rear. The close co-operation of the three arms is the secret of strategical and tactical success. A curious fact, as regards staff duties, and the extreme care that should be taken in drawing up instructions, comes out with respect to Stuart's failures. Lee allowed him to act on his own judgment as to moving round the enemy's rear, although he does not seem to have cordially approved of the idea. But at the same time he ordered him to instruct the commander of the two brigades left behind to watch Hooker, that if the Federals moved northward, he was to watch 'the flank and rear of the army,' moving into the Shenandoah Valley and 'closing upon the rear of the army.' Stuart, in his orders to his subordinates, used the words—'after the army has moved, cross the Potomac and *follow* the army, keeping on its right and rear.' The officer concerned, probably ignorant of the plan of campaign and the distribution of the army corps, did *follow* the army, with what result we know. The instructions he received from Stuart misled him. They attempted to cover all sorts of contingencies. In certain points they lacked precision. No stress was laid on the fact that those two brigades were to act as screens to the army, nor was it anywhere indicated that close contact with the army was above all things essential. In fact, the main point was lost sight of, or obscured by references to less important objects, which might well have been left to the initiative of the recipient. If his judgment could not be trusted, he was not a man to whom the command of a detached force, and so important a duty, should have been assigned.[11]

The second point is the conduct of the great infantry attack on the Federal centre. The Staff, as we have seen, seemed utterly incapable throughout the battle of bringing the efforts of the larger units into timely co-operation, and at the most important crisis of the whole engagement their failure to ensure combination was conspicuous. In the first place, there is no doubt that Lee intended that 30,000 men should have been employed instead of 15,000.[12] In the second place, the supporting brigades on either flank were not well handled; the left brigade was too close to the centre, the right brigade, when Pickett in the centre changed direction a little to the left, moved forward in the original direction, soon found itself isolated, and fell back. In fact there were no supports at hand to confirm success when the crest of the ridge was carried, neither infantry, cavalry, nor artillery.

It is curious that Osman Pasha's splendid attack when he attempted to break out of Plevna was almost an exact reproduction of the Confederate assault at Gettysburg. He had 30,000 men, of which 15,000 formed his reserve. He also had to move over absolutely open ground, and he also was partially successful. Two lines of entrenchments were carried. But when another

11. Stuart's letters, and also Lee's, are quoted in *Battles and Leaders of the Civil War*, ed. Robert Underwood Johnson and Clarence Clough Buel (New York, 1890), III, 252–53.

12. [It is not clear how many troops Lee originally intended to use in the assault on July 3. According to Henderson, "it was never Lee's plan to assault the center only, but both center and flank simultaneously" ("Review of General Longstreet's Book," p. 114). Lee originally intended to make the attack with Longstreet's entire corps, but in the face of Longstreet's opposition he replaced two divisions of Longstreet's corps with a corresponding number of troops from Hill's corps (Freeman, *R. E. Lee*, III, 108).]

effort was required to complete the success, the reserve was not forthcoming. Its passage across the river had been blocked by the carts of the fugitive inhabitants of the town; and nothing was left but surrender. At Chattanooga, again, Grant's most brilliant battle, November 25, 1863, the decisive attack was made on a part of a position which seemed impregnable, by 25,000 men carefully formed up in three lines. I cannot help thinking that these instances show us the necessity of most careful preliminary arrangements when a large mass of troops is sent forward to attack. The whole force should be drawn up with proper intervals and distances. Every commander should have his objective pointed out. No movement should be permitted until every unit is ready to step off at the same moment. Artillery should accompany the attack, prepared, if necessary, to push forward into the fighting line, and cavalry should follow, watching for every opportunity of striking in. Over and over again we read of attacks of this nature which were manifestly unsuccessful because sufficient precautions had not been taken that the whole mass of men to be engaged in the operation should act in close co-operation, because the operation lacked vigour, and because Napoleon's maxim, that in a decisive attack the last man and the last horse should be thrown in, was disregarded.

The explanation of the failure of the Confederate staff is not to be found in the fact that the majority had had very little previous training before the war broke out, many of them being volunteers, pure and simple, or that they were unaccustomed to handle large masses in an attack on a single objective. Two months before, in a far more difficult country, in a dense forest, at Chancellorsville, far more complicated movements had been made by a whole army corps of 25,000 men. I am forced to the conclusion that at Gettysburg Lee's whole army suffered from over-confidence. Face to face with an army they had beaten so often with inferior numbers they relaxed their precautions; and at Chancellorsville the preliminary arrangements for the great attack were made by General 'Stonewall' Jackson, a tactician of the first order, with the utmost deliberation. Not a battalion was allowed to move forward until every man was in his place and every available rifle was thrown into the fight.

The last point I wish to touch upon is the conduct of both the

Federal and Confederate artillery, both before and during Pickett's charge. In the third volume of 'Battles and Leaders of the Civil War' we have descriptions of the battle by the artillery commanders on both sides, and their accounts are a detailed object-lesson in artillery tactics such as is seldom met with.

In the first place, the Federal batteries, although inferior in strength, were never silenced by the Confederate fire, but simply withdrew, in the words of the Chief of Artillery, 'to replenish ammunition and to prepare for the assault which he knew must follow.' On the other hand, we have it on the authority of the Confederate Chief of Artillery that he was completely imposed upon by these tactics. 'He had never,' he says, 'seen the Federals withdraw their guns simply to save them up for the infantry fight.'

Secondly, the latter officer says that the front occupied by the artillery was so long that it was not well studied; the officers of different commands had no opportunity to examine each other's ground for chances to co-operate. Guns which might have enfiladed the Federal batteries playing upon Pickett simply fired straight to the front. In fact, concentration of fire on the tactical point had not been arranged, and dispersion of fire was the result. This brings us to a very curious fact. The two officers in charge of the artillery on either side had served in the same battery in the United States army before the war. The Federal had been the Major, and the Confederate had been placed under him expressly to receive instruction in field artillery. At the final surrender of Lee's army, in April 1865, the two met and the conversation turned on Gettysburg. 'I told him,' writes General Hunt, the Northerner, 'that I was not satisfied with the conduct of the cannonade at that battle, inasmuch as he had not done justice to his instruction: that his fire, instead of being concentrated on the point of attack, as it ought to have been, and as I expected it would be, was scattered over the whole field. He was amused by my criticism and said, "I remembered my lessons at the time, and when the fire became so scattered wondered what you would think of it."' Well, Hunt thought very little of it, for he says that 'most of the enemy's projectiles passed overhead'—he was standing with his own batteries—'the effect being to sweep the open ground in rear—a waste of ammunition, for everything here could seek shelter. . . . In fact, the fire was more dangerous behind the ridge than on its crest.'

The last point of many well worth notice is, that when Pickett advanced, descending into the valley, the Confederate guns reopened over the heads of his troops 'when the lines'—I am quoting the Confederate Chief of Artillery—'had got a couple of hundred yards away, but the enemy's artillery let us alone and fired only at the infantry.'

Here, again, in the action of a large mass of artillery, we have forcibly impressed upon us the importance of careful preliminary arrangements, and the necessity of training officers, when large numbers of batteries are employed, to make co-operation against the tactical objective their first thought.

THE CAMPAIGN IN THE WILDERNESS
OF VIRGINIA, 1864[1]

There is to be found in the correspondence of Napoleon a letter written to an official in France during the great campaign of 1807, which has reference to the theoretical study of the art of war. The Emperor complains that it is very difficult to know what books are useful for the study of military history, and declares that, owing to this difficulty, he had read a great many books which he found quite worthless, and had thus wasted a great deal of time.

It is perhaps a further proof—if further proof were necessary —of the great importance which the greatest of all soldiers assigned to theoretical study, that he should have found time, in the midst of a great army actually confronted by the enemy, to write a letter on such a subject. But it is not my purpose to emphasise the lesson which may be deduced from his words, and to enlarge on the necessity of our making ourselves acquainted with the great campaigns of history. Such a course of study has for its chief end the education of the mental faculties, the strengthening of the intellect, and the development of a capaci-

1. [A lecture to the Military Society of Ireland, January 24, 1894, this article was published in *The Science of War*, chap. xi.]

ty for hard thinking. I can scarcely imagine that it is still necessary to defend the advantages of education; nor is there anyone bold enough nowadays to deny that an active intellect and a capacity for hard thinking are absolutely requisite in any officer who aspires to command troops with honour and success. It is only the uneducated who cry out against education; only the ignorant who are unable to realise the benefits of knowledge; only the man whose ideas of war are absolutely different from those of Napoleon and Wellington, lacking the common-sense with which those great men were so pre-eminently gifted, who dare rail at the study which they considered so essential.

I think that to-day we are all of us quite willing to take the world's most famous soldiers as our masters, and to accept their methods and their teaching as the best means of making and of learning war.

But Wellington and Napoleon are not the only masters of war, and I should like to bring to the notice of our rising soldiers a very great campaign which has by no means attracted the attention it deserves, yet which is full of instruction for officers of all ranks, and, in my humble opinion, gives a better clue to the fighting of the future than any other which history records. In May 1864, when the campaign began, the Americans had been fighting for just three years. Their armies, which had to be improvised on the spot, out of a civilian population, absolutely innocent of all military knowledge, were not very good for the first year or so. They were certainly not equal to regular troops. It is hardly possible, when we consider the disadvantages under which they laboured, that it could have been otherwise. But three years of active service told their tale. General Sherman, a man whose ability and honesty none can deny, has written that after 1863—that is, in the year of the Wilderness campaign—they were equal to any European troops. I see little reason to doubt the accuracy of this observation, and I believe, moreover, that in very many respects the American armies of which he spoke were superior and more advanced in military knowledge than even the Germans in 1870. The American regular officers who filled the higher grades were remarkably well-educated and well-trained soldiers before the war began, and it would have been strange if three years' experience in handling huge masses of men, of incessant fighting against very gallant enemies in a very difficult country, had not stimulated the acute American intellect, already well cultivated, to evolve strategical and tactical methods admirably adapted to the needs of modern warfare.

What these methods were I shall try to make clear, and I think that some day the majority will be induced to agree with my high estimate of the value of this campaign as a clue to the fighting of the future. The American armies were composed of volunteers, with a small leaven of regular officers, who filled the higher commands and the principal appointments on the staff. Now I do not think I am predicting impossibilities when I say that armies somewhat similar in constitution may at some future date have to be handled by ourselves. England has before now been drawn willy-nilly into continental wars; she has before now had to engage in a life-and-death struggle with the Great Powers, and the early part of the nineteenth century saw her troops engaged, not only on the mainland of Europe, but in almost every important island in the Mediterranean, and, what is perhaps more to the point, in almost every single colony or outlying dependency in possession of her enemies. In the great French war, although transport was far more difficult than it is to-day, there were few parts of the globe to which the English navy did not convey English troops; and a list of the various countries and islands which were captured, occupied, and garrisoned by English soldiers is very suggestive reading. The very names on our regimental colours remind us that at every point of a hostile or friendly State which can be reached by sea those colours have been planted; and history tells us with what extraordinary effect the combined naval and military force of England, often insignificant in numbers, but backed up by a long purse, have struck at the resources, the commerce, and the prestige of her most formidable enemies.

History repeats itself. There is no sign whatever, despite long years of peace, that the prospect of our being drawn into a great European conflict is more remote than heretofore. Increased prosperity, greater wealth, and wider interests can scarcely be considered as security in themselves against attack. It is true that in the navy we have our first line of defence, but this very title proves its weakness. The navy is a defensive force, pure and simple, and without the assistance of the army it is passive; it can ward off the blow; but it cannot return it, and if our efforts were confined to naval operations, the counter-stroke, the soul of the defence, would be impossible. We could scarcely hope either to annihilate or to exhaust our enemy. It is possible my judgment may be at fault—I stand open to correction—but as yet I see no cause to believe that in any future European struggle in which we shall be engaged our traditions will be forgotten, and that British troops will not be despatched to occupy those extremities of the enemy's possessions which the command of the sea lays open to our attack. I cannot imagine that our duties will be limited to garrisoning ports and coaling stations, and I can easily conceive a second Peninsular or Crimean campaign. And when we consider the large resources which we have now at our disposal, the enormous reserve which the Volunteer force of Great Britain and the colonies provides, it is still more difficult to imagine that this reserve will prefer to remain idle when the honour of the country is at stake.

If I see in the future an English general at the head of an army far larger than that which drained the life-blood of Napoleon's empire in the Peninsula, if I see our colours flying over even a wider area than in the year which preceded Waterloo, you may think that I am over-sanguine; but to my mind the possibility exists, and with it the probability that the forces which are employed upon the counter-stroke will be constituted, at least in part, as were the armies of the American Civil War. Our men will not all be regulars. They will come straight from civil life, and to civil life they will return. The habits and prejudices of civil life will have to be considered in their discipline and instruction, and officers will have to recognise that troops without the traditions, instincts, and training of regular soldiers, require a handling different from that which they have been accustomed to employ. To my mind this is one of the most important lessons to be learned from the American War by English soldiers. Some of the American officers could get as much out of the volunteers as out of veteran troops. Others, who did not understand their peculiar prejudices, failed to acquire their confidence, and, despite their ability, failed in every operation they undertook. With regulars they would probably have been successful; with volunteers they fell from disaster to disaster. It is possible that all will not agree with me. Some may consider that the system of command adopted for the regular army is applicable to all troops who wear a uniform. But a close study of the American campaigns has forced upon me the conviction that it is not sufficient to bring volunteers under military law. The rules and regulations which govern the regular army are doubtless enough to ensure their obedience and subordination, but something more is required to secure their confidence, and to make them reliable under circumstances of danger, difficulty, and hardship. What this is may be learned from the lives of the American generals, Lee, Grant, Sherman, Sheridan, Hancock, and, perhaps above all, Stonewall Jackson. The following is an extract taken from an article on Hancock, one of the most successful corps commanders in the Wilderness campaign.

'He never sneered at the volunteers. He made them feel, by his evident respect, his hearty greeting, his warm approval of everything they did well, that he regarded them just as fully, just as truly, soldiers of the United States as if they belonged to his own old regiment. Such was the spirit in which Hancock met his new command. We know with what assiduity, patience, and good feeling, what almost pathetic eagerness to learn and to imitate, the Volunteers of 1861 sought to fit themselves to take their part in the great struggle. He saw that it was of extreme importance to promote the self-respect and self-confidence of the volunteer regiments, to lead them to think that they could do anything, and were the equals of anybody. But Hancock was a man of sound common sense, who understood human nature thoroughly, and was therefore fit for high command. He was not a mere drill-sergeant, not a mere fighting animal, and not a mixture of officialism and routine.'

This is not the only point on which English soldiers can draw instruction from a study of the war. The command of the sea, and combined military and naval operations, played throughout a most important part, and in the Wilderness campaign the strategy of the attacking side depended on the same facilities for changing the bases of supply and the lines of operations as were made use of in the Peninsula, in 1854, and in 1882.[2] In this respect, at least, the operations of the Federal army were the counterpart of many English campaigns. Again, the country over which the troops moved and fought was difficult in the extreme. The maps available were few and bad. Virginia, the theatre of war, was thinly populated—not half opened up. A great part of the State was covered with primeval forest. There were immense tracts of swamp and jungle which were terra incognita to all but a few farmers and their negro slaves. The roads were as scarce and indifferent as the maps. The country produced but little in the way of supplies; and the invaders, when they crossed the border, had the very difficulties to face which so often confront English troops, engaged in rounding off the corners of the Empire by annexing some considerable tracts of savage territory. The organisation of the auxiliary departments, the supply, the medical, and the reconnoitring, which enabled the Americans to overcome those difficulties, afford valuable suggestions to ourselves.

I may also notice, though the same observation applies to the study of any campaign whatever, that there is much instruction to be gained on two points on which text-books and field-exercises are necessarily silent, and which are yet of far more importance than strategical dispositions or formations. The first of these is that almost indefinable force which Napoleon declared was as to the physical, that is to numbers, armament, and physique, as three to one. Any general who ever made war successfully relied far more on the moral effects of his manoeuvres than on the mere fighting qualities of his troops; and it may be said with absolute truth, that it was because he understood the immense power of moral influences that he was successful. But as it is the most important, so this factor in war is the very hardest to teach. Still it can be taught, or rather it can be learned, and I cannot help thinking that it is to this that Napoleon referred when he said that reading and re-reading the campaigns of the greatest captains was the best means of learning the art of war. I should find it by no means an unpleasant task to discuss this subject at length, but I can do no

2. [In 1854, the British, French, and Turkish armies invaded the Crimean peninsula; in 1882, the British intervened in Egypt to suppress an uprising headed by Arabi Pasha.]

more here than to advise young officers, whenever they take up a book of military history, to keep this factor always before their minds, to note every instance in which it exerted an effect, to take to heart the way in which it was employed, and to remember that it is to a thorough comprehension of its value rather than to mechanical aids, such as formations and fortifications, that the greatest captains owed their victories. The second point to which I refer is the individual character of the commander. I do not mean to say that we can all of us, by merely realising the mixture of prudence and audacity, the iron will, the invincible determination, the dogged perseverance, and the incessant application of, for instance, Wellington or Moltke, become Wellingtons or Moltkes ourselves. We came into the world endowed with certain mental and moral attributes; we have, all of us, our weak points, some perhaps have strong ones, but we were not created equal in this respect or in any other. Nor is the moulding of our character altogether in our hands. But it is useless to deny that, as in some degree at least we are masters of our own fate, we may be masters in some degree of our own natures. Example is a potent force in this world. We may never reach the ideal after which we strive, but it is within our power to approach it; and the effort to acquire the qualities which have distinguished great soldiers will not be a barren one.

The memories of what they did and what they dared may inspire us some day to imitate them, however feebly; and even a weak imitation may be superior to the working of natural impulse. In military history the very highest ideals may be found; and here again I would advise students of campaigns to mark the influence of the character of great soldiers on difficult operations, and to learn how determination, perseverance, and the fixed resolve to conquer, have enabled them to triumph over obstacles before which men of weaker fibre would have turned aside. To keep these points always before our minds, the influence of *moral* and the influence of individual character, is the true way of studying military history.

With these observations I come to the campaign itself, and I must now explain the general situation and describe the theatre of war. The Civil War, as I have already said, had, at the beginning of May 1864, been going on for three years. The respective capitals of the United States and the Confederacy were Washington and Richmond. Richmond had been the great objective of all the fighting throughout the war. To capture Richmond was, in the opinion of the Northerners, to break the back of the Rebellion and to end the conflict; and their efforts throughout had been directed to this end. During the preceding three years they had made no less than five attempts to reach the Southern capital. Each one of these attacks had been beaten back.

In May 1864, the United States Government once more resolved to attempt the seemingly hopeless task. The Northern army (the Army of the Potomac) was composed of the same troops that had been engaged in these various expeditions. The Southern army (the Army of Northern Virginia) was the army which had beaten them back to Washington. Their respective strength at this time was as follows:—The Army of Northern Virginia, covering Richmond, consisted of three army corps, two cavalry divisions, and 224 guns, giving a total of 62,000 officers and men. The Army of the Potomac mustered 130,000, divided into five army corps, four divisions of cavalry, and 316 guns.

At the head of the Confederate army was General Lee, undoubtedly one of the greatest, if not the greatest, soldier who ever spoke the English tongue. He had been in command of the Army of Northern Virginia since June 1862—that is for two years, two years of incessant fighting and of numerous battles—and he possessed in a very extreme degree the confidence of his officers and the affection of his men. The Federal army, during this eventful period, had been commanded by several different generals. The Government elected the best general they could find at the beginning of the war; when he was beaten they relieved him and sent another. Five generals in succession had held the chief command. In 1864 came the turn of Grant. Grant had hitherto been fighting in the west far away on the Mississippi, where he had won some extraordinary victories, and had displayed great ability both as strategist and tactician. As Commander-in-Chief of the whole United States army, the position to which he was now appointed, he had to devise a plan to capture Richmond, and to this end no less than four armies were set in motion. Whilst holding in his own hands the control of the campaign, he established his headquarters with the Army of the Potomac, the pivot of the whole scheme of invasion; for before that army lay the main force of the Confederates, its old rival, the Army of Northern Virginia; and it is well to remember, in order to appreciate Grant's difficulties and his strength of character, that with strange troops he had now to encounter a most formidable adversary, and that those troops had far more dread of Lee than confidence in himself.

At the beginning of May Grant decided to march on Richmond. His headquarters were at Culpeper in Virginia, for the Federals had mastered a certain portion of that State, and his troops, generally speaking, were massed round that town. Lee with his 62,000 men stood opposite, and the river Rapidan, a wide and deep stream, ran between the hostile camps. The Confederate headquarters were at Orange Court House, and the troops extended along the river bank in a strongly entrenched position. On the right flank of the line there ran a stream called Mine Run, and along this stream was a return entrenchment, striking due south from the river. The dispositions of the leaders raise an interesting question. Lee had to cover Richmond. The Federal army was posted at Culpeper, so he took up a position opposite to them and entrenched himself. His right flank was very strongly guarded by the return entrenchment, and his left flank was also strong by reason of the country; he had little fear that he would be seriously attacked in that quarter. Grant, when he reached Culpeper and took over command, found his opponent directly in front of him, covered by his formidable lines, and to all appearances barring the way to Richmond. He at once came to the conclusion that it was no good attacking the Confederate position; there was not only the river to be crossed, but there were the entrenchments to be carried. Should he move round and try to turn Lee's left? The railway which runs south from Culpeper afforded a line of supply which would have greatly facilitated this operation. But if he worked by that flank Lee would fall back to some new defensive position, still covering Richmond, and the Federals would find no opportunity of fighting him at a disadvantage. It is important to note carefully, as a clue to the operations, that Grant was not aiming to avoid Lee and then seize Richmond. That would scarcely have been a judicious plan. Richmond was fortified, and he could not have held the town with the Confederate army intact behind him and cutting his communication with the north. His intention was to crush Lee first, and then deal with Richmond at his leisure; and in order to crush Lee with certainty, he wished to catch him at a disadvantage; *i.e.* to attack the Army of Northern Virginia in the open, on ground where it would have no time to entrench, or, by intervening between it and Richmond, to compel Lee with his inferior numbers to attack the army of the Potomac. Putting the first two lines aside as impracticable or unpromising he only had a third left, and he determined to move south past the Confederate right.

A glance at the map [Map I] will show that his line of supply, the Orange and Alexandria railway, ran past the Confederate left, and in selecting a line of operation by the opposite flank, he would have to abandon his communications. This was even a more momentous matter in Virginia than elsewhere, for there were no supplies whatever to be procured in the country. The question of provisions was a most difficult one, but it had no influence on his determination. He still held to the plan he thought most promising of success, although, in order to be free for protracted movements, the army would have to carry ten days' rations for man and horse. These ten days' rations for 130,000 men, together with ammunition and medical supplies, required about 5,000 wagons, a very great encumbrance to an army, especially in a country where the roads were few and bad. It is evident that Grant had no easy task. Remember that before he could pass Lee's right flank he had to cross the Rapidan, and that his movement, which should partake of the character of a surprise, was bound to be hampered by his enormous train. He resolved to march under the cover of the darkness. His orders were issued on May 2, and at midnight on the 3rd the troops started. At dawn they reached the river, the cavalry leading, laid five bridges, and by the night of the 4th nearly the whole army and a portion of the train had passed. It was certainly a successful operation to get these enormous numbers over the river safely.

We now come to the question why Lee, who had to cover

Richmond, made no attempt either to prevent Grant's passage, or to put himself in his way when he had got across? This is a most interesting point in the campaign, and it gives some idea of Lee's ability and daring. He knew well enough that Grant would endeavour to turn his right. He had told his generals several days before exactly what would happen, yet he made no attempt to stop his enemy crossing the Rapidan. He did not allow half of them to get across, then fall upon them and send them back defeated. He let the whole army make the passage without the slightest molestation; and remember he had only half the number of men that Grant had—62,000 against 130,000. But south of the river was a tract of peculiar country, a district which was simply a jungle, significantly called the 'Wilderness of Virginia.' It extended about ten miles south from the Rapidan, nearly as far as Spotsylvania Court House, and through this jungle lay the Federal line of march. Before Grant could get out into the open country he had to pass through the Wilderness. The Confederates, nearly all of them Virginians, knew this district well. Lee had already fought a successful battle against overwhelming odds in those very thickets, and he determined to let Grant entangle himself in the Wilderness and there attack him. In that most intricate country where artillery could not be used, where men familiar with the paths and clearings would have a good advantage over far superior numbers, he would throw his 62,000 men on Grant's 130,000. Whatever may be said of his judgment, everyone must admire his boldness; and this was the plan he had in view when he allowed Grant to push quietly across the river and bring his enormous impedimenta with him. When he found that the Federal army was well over, he marched east from Orange Court House and attacked it in the Wilderness. Nothing would serve him but to annihilate the whole. The Confederates, however, had a long day's march to make before they could reach the field of battle their leader had selected; so, after crossing the Rapidan, Grant had twenty-four hours to himself, twenty-four hours in which to place his army across Lee's road to Richmond. His cavalry, scouting to front and flank, down the forest roads, found no signs of the enemy; there was nothing to prevent a rapid movement south; the Confederate commander had been apparently taken unawares; and, if he had moved at all, had merely occupied the return entrenchments along Mine Run, a position very strong in itself, and on the flank of the Federal line of march.

This position it would not be difficult to turn. There was comparatively open country to the south, where troops could manoeuvre with ease, and the superior numbers of the Federals could be made full use of. There only remained to get clear of the Wilderness. This could not be done on May 4. The infantry and cavalry could have easily made the necessary march, but the 5,000 wagons took nearly thirty hours to cross the river, and the troops had to remain encamped in the jungle to protect the train. But next morning, the 5th, although the whole of the train had not yet crossed, the Federals struck south. Scarcely had they started on the march when Lee's columns dashed against their flank, and the battle of the Wilderness began.

To go into the details of this battle would take much too long, but it is interesting as an example of wood-fighting on a most extended scale. The armies fought for two days in the jungle. The Federals, however, were not beaten; their losses were very heavy, but they just managed to hold their own. The troops fought well, and they brought to their aid one of those new methods of warfare which the Americans had invented. Both sides suddenly found themselves within a few miles of the enemy. I need not say that in this very thickly wooded country the cavalry found themselves at a very great disadvantage; they could get very little information. But the infantry took good care of itself. Directly any brigade or division found that the enemy was coming up, it sent out scouts to reconnoitre and immediately entrenched. There was no waiting for orders. If the general did not give the order, the battalion or company commanders acted for themselves, and it is even said that the men, directly they halted, threw up shelter without waiting for their superiors to give the word. The entrenchments were strong enough; and in this wooded country they were easily constructed. There were a great many expert axe-men in the armies, and trees were soon felled, or the fallen timber gathered. A pile of logs and branches made a good foundation, over these the men threw a little earth, and a parapet was soon constructed that was bullet-proof at least. With both sides entrenched, the course of this battle was simple in the extreme. One side came out from its entrenchments and attacked, got beaten and retired; the enemy followed in pursuit, but was brought up in turn by the entrenchments. In this thick wood manoeuvring was almost impossible; what little took place was undertaken by the Confederates, who knew the ground. The troops were obliged to use the roads whenever they made a movement in any force; and it is an interesting point to note that there was a great deal of marching by the compass. The forest was so thick that this was the only way the battalions could keep in the right direction. The losses in this battle were very great. The Federals, during these two days, lost 15,300 men and officers; the Confederates 11,000. Bearing in mind the supreme importance of individual character I may call attention to the conduct of the rival commanders. It is impossible not to recognise Lee's audacity. Although he was doubly outnumbered he allowed the Federals to cross the river at their leisure; he made no attempt whatever to interfere with them until they were involved in the difficult country in which he wished to find them. It is true he did not defeat them, but he dealt them so staggering a blow, and inflicted such heavy losses, that he might well anticipate that retreat would be their only thought. But in Grant he had a foe of more than ordinary tenacity. The Army of the Potomac had been defeated over and over again, and it is not too much to say that every general in the Federal army had hitherto considered himself inferior to Lee. With some of them it was like the old days upon the Border, when the English mothers used to stop their crying children with the name of the Black Douglas.[3] The mere mention of Lee's name to the officers of the stamp of Hooker and Burnside seems to have been enough. They were paralysed at once.

Now, here was Grant, a stranger to his troops, face to face with the hero of the war, the man before whom so many generals had gone down. He had fought him for two days in the Wilderness, and if he had escaped defeat he had lost a great many more

3. [Sir James of Douglas, a Scottish nobleman, won the dreaded name of "Black Douglas" for his frequent raids on the English border during the first decades of the fourteenth century.]

men than Lee, and the fighting all through had certainly not been in the Federal favour. The morning after the battle they brought in a list of losses—15,000 men—and the enemy was still there: still there and not retreating! Grant had to decide what to do; it was little use attacking the enemy in his entrenchments; there seemed no hope of success, and the army would not have been surprised had he followed the example of his predecessors and retreated. But despite his losses, despite the demoralisation of his troops, despite the fact that he had not won an inch of ground, he determined to move forward, to follow out his original plan, and, if possible, to cut Lee off from Richmond, or at all events to force him to battle in a less impregnable position than the one he now held. This was the turning point in the campaign. In so deciding he had to face the difficulty as regarded communications. He had only seven days' provisions left, and there were all the sick and wounded to be sent to the rear. But the Federals had command of the sea. Moreover, several great water-ways run up into the heart of Virginia. There is the Rappahannock, and north of the Rappahannock is another and a larger river—the Potomac—which runs past Washington and the Northern Border. Both these rivers are navigable, and by means of his command of the waterways Grant was able to change his base. He shifted it at once from the Orange and Alexandria Railway to Fredericksburg on the Rappahannock. On May 7 the Federals marched south, and again they marched at midnight. Grant's idea was to intervene between Lee and Richmond, to entrench himself and compel his enemy to attack. He had quite realised the value of entrenchments.

And now came a most curious race, in which Grant had a little the worst of the luck. He made all his preparations to get off as secretly as possible. He sent his trains away in the afternoon, and the troops were not to move until darkness had set in. Lee had an idea that something was going on. He expected that Grant, like the other Federal generals, would fall back upon his base, but he had an idea at the same time that the Federal general might move on Spotsylvania Court House. The shortest road to Richmond ran past the Court House, so that this insignificant village was of the first importance. He therefore made preparations to meet all eventualities: and at the same time that Grant gave orders for his troops to march at midnight, Lee gave orders that a road was to be cut through the woods in the direction of Spotsylvania, so that one of his army corps could get there without delay. But this corps was not directed to march until the next morning. It was to move at 3 A.M.; Grant intended to start at midnight, and the Federal route was by very little the longer of the two. But, luckily for the Confederates, the army corps which was instructed to start at 3 A.M. did not wait so long. The neighbouring woods had been set on fire by the battle, and the general commanding the corps took upon himself to modify his orders. He wanted to escape from the blazing forest, so, instead of waiting till 3 A.M., he marched an hour before midnight. Whilst the infantry were marching through the night on Spotsylvania the Federals had sent on their cavalry to seize the Court House. But Lee had done exactly the same thing, and when the Federals arrived almost in sight of the village they found the way blocked by the Confederate horsemen.

This incident shows the value of cavalry who can fight dismounted. The Confederates had entrenched themselves all along the front, and the entrenchments were manned with rifles. Although these rifles were only held by cavalry soldiers, the Southerners managed to keep a much superior force in check until their infantry came up, and General Lee's army was the first concentrated round Spotsylvania Court House. When Grant reached the field he was much disappointed to find that he had been outmanoeuvred, that his midnight march had been no good, and that he was again confronted by lines of breast-works.

On the next day, May 9th, began the great battle of Spotsylvania—at least it is called a battle, but it was really a series of engagements that continued for about nine days. The sketches show how skilfully Lee had made his dispositions. He took up a position between the two streams which are called respectively the Po and Ny; his front was exactly adapted to the numbers he had at his disposal; in order to turn the position his adversary would have to cross one of the streams, and so divide his army, giving him an opportunity of dealing with him in detail, and his line was far stronger than that which he had held in the Wilderness. The country was still very thickly wooded—the Federals had still to face their old difficulty of finding out where the enemy was and in what direction his entrenchments ran. The first two days were occupied in reconnaissance. Reconnaissances, as we read about them in text-books, are always executed by the cavalry. The worst of it was that, although the Federals had plenty of cavalry, they were absolutely of no use at all in such a country; and so information had to be obtained by simply sending out brigades of infantry to stir up the enemy, and to see if he was in position at such and such a point. Reconnaissances in force were therefore the only means by which the Federals could find out anything about the enemy; and it is worth remembering, because reconnaissances in force are not operations with which we have much to do, and a good deal can be learned from these campaigns as to the manner in which they should be carried out. On May 10th the Federals had gathered sufficient information as to the enemy's position. The first thing they did was to send an army corps across the Po to see whether they could turn Lee's left; but Lee was entrenched so strongly behind the stream that attack was not permitted, and the corps was withdrawn after beating back a counter-stroke. This was on the morning of the 10th. By the evening they had found that at a certain point on the opposite flank the Confederate line was more accessible, and Grant ordered that while one corps kept the Confederates employed, a strong attack should be made on the weak point. The formation adopted for the attack is interesting; the same principle was observed which obtains to-day on the Continent, and which is advocated in our own Drill-book. I say the principle only; I do not mean to convey the impression that the Federal troops observed the same intervals and distances that are now laid down. Three divisions were employed: one, on the right, was formed in two lines; two-deep lines with a few skirmishers out in front at about 200 paces distance. On the left there was the same formation, a second division was formed in the same way; but in the centre there was a heavy column of twelve battalions formed in four lines, at 100 paces distance.

The idea was, that the right and left wings should attract

the enemy's attention and attack first, and that the central column, massed under cover, should rush the entrenchments. It is well to remember that breech-loaders were not used in America except by the cavalry;[4] but the infantry had rifles, and very good rifles, for they could kill at more than 1,000 yards. About 300, or at most 400, yards was the effective range, but for all that they were very useful weapons although they were muzzle-loaders. This attack was perfectly successful. It was prepared by thirty pieces of artillery, and the central column managed, by making use of the shelter of the wood, to get close to the enemy's works before it was observed. The attack of the two wings engaged the attention of the Confederates; when the word to advance was given, the whole twelve battalions moved off as one man, charged the breast-works, swept clean over them, took 1,200 prisoners, captured twenty guns, and carried a second line of entrenchments in rear. But the Confederates had reserve brigades close at hand. These made a determined counter-stroke, and the Federals, in all the confidence of a successful attack, were driven out nearly as quickly as they got in. The men were exhausted; they had made a long charge, the fighting within the works had been very heavy, *and there were no supports.*

There is a useful lesson emphasised here. These great masses of men, in several lines, one behind the other, as has been shown over and over again, if the ground is at all favourable, and the propitious moment seized, will go through anything, but if you want to keep what you have won you must have strong reserves behind. The same thing occurred at the Alma, where the great redoubt was carried without any great difficulty; but when the Russian columns came forward to the counter-stroke, the men looked back, and seeing no supports in rear, they streamed away. Much the same thing occurred at Spotsylvania on May 10th. I may add that, despite their deep formation, the Federals lost but few men until they were attacked in turn; the actual charge—the storm of the entrenchments—was not at all a costly proceeding.

4. ["Breechloaders—and repeaters at that—were used by a few infantrymen in the latter part of the war and with great effectiveness. . . . But they were not an item of general issue to foot soldiers and their use was restricted largely to cavalrymen" (Bell Irvin Wiley, *The Life of Billy Yank* [Indianapolis, 1952], p. 63).]

On the 11th there was more reconnaissance, and the same evening General Grant determined on an attack on a larger scale. The central point of Lee's entrenchments, salient to the remainder of his lines, was believed to be the weakest part of the position, and during the night 20,000 men were massed against it. The formation was similar to that which had been partially successful on the 10th. There was one division on the right in two lines; a second in the centre, with a third in rear, but the battalions of these two divisions, instead of being in line, were formed in column, in fact they were in line of masses, and each battalion was in column of double companies. Perhaps the most interesting point in this attack was the manoeuvring which took place before it. The whole army closed to its left, and the corps that made the grand attack was brought into position by a night march of some four or five miles, forming up outside their own entrenchments at 1,200 yards from the Confederate lines. Twenty thousand men were thus massed ready to attack at daybreak; and that they were able to march through dense woods where maps showed nothing, where the tracks were only known to the few guides, and to form for attack in the darkness with silence and precision show that staff duties in the Northern army were by this time thoroughly understood.

At half-past four on the morning of the 12th this enormous mass of men rushed forward, swept over the open ground in face of a heavy fire, tore away the abattis, and stormed the parapet. Holding the entrenchments was the best division in the Confederate army, but nearly the whole were captured, together with twenty guns, two general officers, and seven stands of colours. Nor were the Federals satisfied with this first success. The men pressed forward, and sweeping everything before them, drove the thin end of the wedge right into the Confederate lines. But Lee, recognising the weakness of the salient, had caused a return entrenchment, or rather another line of entrenchments, to be constructed about half a mile in rear. By this second line the Federals were suddenly brought up. The confusion was very great, the battalions had intermingled in the excitement of the charge, and the officers could neither make their orders heard nor form their men for another rush.

Lee threw in his reserves. He made a tremendous counterstroke. Every single battalion he could collect was ordered to attack; and the vigour of the blow was such that the whole of these 20,000 men were driven back beyond the first line of entrenchments, and the Confederates recaptured their first position.

The fighting that followed furnishes one of the most extraordinary stories in the annals of war.[5] The infantry on both sides lay for the whole day with the parapets between them, in many places not more than ten or twelve paces distant. But in the end the Federals had to retire; they had found it impossible to break through the Confederate line. We may notice how nearly this great attack came to a complete success, and that the cause of its ultimate defeat was that in the excitement of the attack the second and third lines, instead of keeping their respective distances, closed in upon the first. I believe it is the experience of many officers who have been engaged in similar attacks that it is very difficult indeed to keep the men in hand, and that second and third lines invariably act as did the Federals. The column on which they principally depended, as soon as the first success was won, became a confused mass of men over which officers and non-commissioned officers had no control whatever, and when these men struck the second entrenchment they were merely a mob. It was said afterwards, by officers who had taken part in the fight, that the distance between the lines ought to have been very much increased, and that the second and third lines ought to have waited until they saw they were wanted, and not to have reinforced till then.

In every country in Europe, France, Germany, Austria, Italy, so far as the principle goes, these deep formations—not of course in columns—have been generally adopted. The principle is that you mass opposite one point a great wedge which you intend to drive into the enemy's line, and that this wedge is composed of several lines one behind the other at such distance as may best suit the ground and the situation. The same formation, I need hardly repeat, is advocated in our own Drill-book. But it is a great point to remember that you should have a force behind this wedge in order to confirm success when you have broken in, for whatever may be the discipline of the

---

5. [This was the well-known "Bloody Angle."]

troops it is impossible that confusion and intermingling of units can be avoided.

After the 12th General Grant determined to try at a new point. He was not done with yet; in the great attack he had inflicted heavier loss on the Confederates than they had on him, and his men, although they were beaten off, had fought so well and had come so near victory that they were quite ready for another effort. A movement by the left promised an opportunity of attacking Lee's right before it could be reinforced or his entrenchments extended. Grant therefore moved nearly the whole of his army by a night march of several miles to a line opposite to and outflanking the Confederate right. But here again he had the worst of the luck. During the night the rain fell in torrents. The roads were knee deep in mud. It was so dark that even the torches did nothing to make it brighter, and the men struggled wearily along at a very slow pace and with many halts. When day broke the advanced guards had reached the appointed rendezvous, but the columns in rear were so strung out and scattered, and the troops so utterly exhausted, that all idea of attack had to be abandoned.

This was unfortunate for Grant, as General Lee, who had no information of this new move, had very few troops on his right flank. If the roads had been dry it is exceedingly probable that the Confederate entrenchments would have been stormed. We have now reached the 14th. For the next three days Grant remained in position opposite Lee's right, resting his men, and receiving reinforcements; then he made another night march, returning to the scene of the great attack. Grant's idea was that he had been facing Lee's right for a long time, and that the Confederate general, expecting an attack on that flank, would probably have thinned his line in the centre. But Lee had done no such thing. He had a suspicion that his enemy might manoeuvre once more, as he had done already, and he not only held his centre in force, but had strengthened it by abattis and artillery. So when Grant had marched round, and once more attacked the salient, he got well beaten; the position was a great deal too strong to be attacked. This was the end of the fighting at Spotsylvania. The Federals had lost 17,000 men, the Confederates about 12,000.

On the 21st Grant determined to strike boldly round the Confederate right, and if possible to force Lee to attack. The operations which followed are too complicated to describe here. The main fact is that Lee found out, by means of his cavalry, what Grant was doing, that he refused to fall into the trap which his opponent had laid, and, slipping quietly away, still making use of his interior lines, interposed between the Federals and Richmond behind the North Anna river; there he had two bridges opposite his right, a ford opposite his centre, and another ford two and a half miles distant beyond his left. I do not think that he believed that Grant would come over and attack him. He rather believed that he would move off once more past his right flank. When Grant, however, reached the river, and found Lee behind it, he determined to try the strength of this new position. He, therefore, ordered one of his army corps to cross the ford beyond the enemy's left, reconnoitre the Confederate position, and if there was any prospect of success to report at once. This corps crossed the bridge [ford], and, as usual, immediately threw up a line of entrenchments. Now, Lee had, hitherto, been holding his own against the Federals with much success, but he had not yet defeated them. When he saw one corps cross the river, more or less isolated from the remainder, he recognised the opportunity he desired, and he ordered an immediate attack. But he was not present himself during the engagement; unfortunately for the Confederates he was lying sick in his tent. However, he sent one of his best generals in command of the attacking corps, but the counter-stroke was unsuccessful. The Federals had entrenched, and when the Confederates came on and assaulted the breast-works, they found to their cost what a difficult business such an attack was, and the defence once more prevailed. Grant reinforced this corps by a second, and moved a third over the bridge opposite Lee's right. As the situation now stood, he had rather the advantage. One corps was still beyond the river, opposite Lee's centre, and if he could have thrown this corps over the ford in front of it, he would have had everything in his own hands, and have been able to crush the Confederates. He was much superior in numbers; his troops across the river were strongly entrenched, and he had no reason to fear attack.

Lee now put into practice a very curious manoeuvre. His

army was more or less separated. The corps on the left was three miles distant from those which held the right and centre, so it was possible that he might be beaten on one wing before his reserves could reinforce it. His line in fact was dangerously extended. He got out of his difficulty in this way:—he shut up his line like one closes an umbrella; the line had originally been almost straight, it now assumed the shape shown in the map. [Map VII.] His whole force was now massed in a space not more than two and a half miles broad, and his enemy was not only widely separated, but would have to cross the river to reinforce one wing from the other. He could reinforce a point attacked in one-third of the time that Grant could reinforce at the same point. Grant was completely nonplussed by this manoeuvre, in fact his only idea was to get out of his uncomfortable situation as fast as possible. He found that he had two corps on one wing, one corps on the other, separated by a wide interval and by the river. It was evident that nearly the whole Confederate army might fall either on one or on the other. As a matter of fact here was a very great opportunity—so say the critics—which Lee might have seized, and which, if he had been himself, he probably would have seized; but as fortune would have it, when General Grant was entangling himself in this most awkward position, Lee, as I have said, was sick in his tent. On the night of the 26th, Grant got out of his difficulties by recrossing the river under cover of the darkness, and once more he moved round Lee's right. Lee followed suit as before, and the two armies eventually came into contact at Cold Harbor, and here was fought the last battle of the Wilderness campaign. General Grant advanced, hoping that he would find Lee getting into position, but he found instead that the Confederates were already entrenched with their flanks secured by streams, and that there was no chance of catching them at a disadvantage. And then at last he seems to have lost his temper. There was no manoeuvring at Cold Harbor as there had been at Spotsylvania; there was no massing against one particular point; but the army moved straight against the Confederate front, and the order was given, 'the whole line will attack.'

There was no attempt at any formation beyond drawing up the army corps each in two lines. The artillery was ordered to

MAP VII

The Campaign in the Wilderness

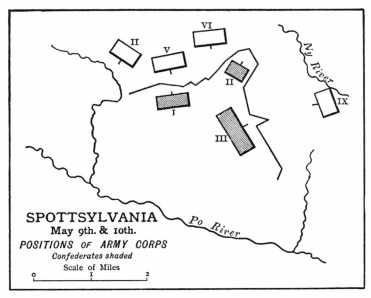

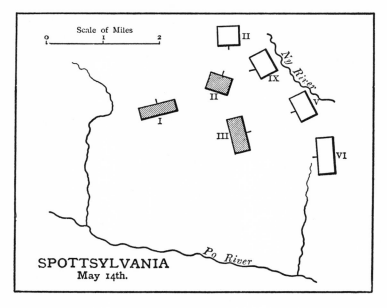

MAP VII—*Continued*

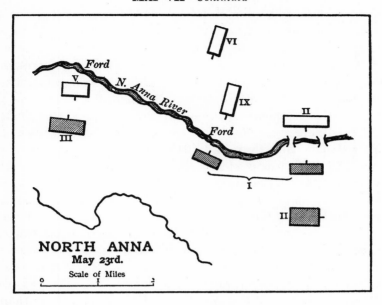

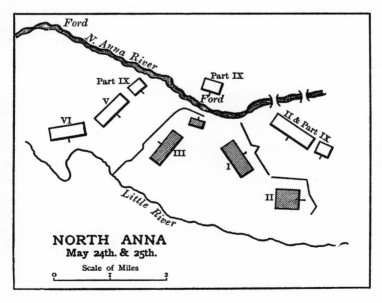

do what it could in the way of bombardment, but that was very little; and when the attack was made it was driven back in little more than an hour with a loss of 12,000 men. Grant sent a fresh order that the attack was to be renewed, but the men lay still and would not move. The American soldiers had sometimes a way of their own of expressing what they thought of their general, and this time they showed him that such attacks against entrenched positions were absolutely impossible.

This battle took place on June 3rd; it was confined to a single attack, and here again the Confederates made no attempt at a counter-stroke. But they had little opportunity. Before the attack was made the Federals had constructed long lines of entrenchments, and Lee and his generals had found out by experience that it was no good attacking these hasty fortifications. During these operations Grant had again changed his base. Every time he moved by his left flank and tried to get round Lee, he shifted his base along the water-ways. First of all be began with the base on the railway; then he went to Fredericksburg, then to Port Royal; next to the White House, and eventually to the River James. He changed his base no less than five times; his army was always well supplied, even his enormous numbers of wounded were carried straight away to the base and thence to Washington, without any difficulty; and he had no obstacles whatever to fight against as regards either feeding his men or keeping up the supply of ammunition.

The end of the campaign, so far as we are concerned, is the passage of the last great river, the James. The James, below Richmond, is as broad as the Danube near Vienna; a very difficult obstacle indeed; and it is curious to find that, notwithstanding this difficult obstacle, Grant not only carried the greater part of his army over before Lee was made aware of his movements, but that he very nearly defeated a portion of Lee's army, and captured a section of the earthworks which defended Richmond, from the south. After Cold Harbor, Grant threw all his cavalry towards Richmond along the White House Railway. They came in contact with the Confederates, and the Confederates could not discern what was going on behind this screen. Meanwhile all the infantry of the army moved down to the James, and made the passage. Grant had now determined to attack Richmond from the south, cutting the communications of the capital with the rest of the Confederacy, and in making his flank march he most certainly outmanoeuvred Lee. It was only the slackness of one of his subordinates that saved the Confederate army, not indeed from defeat, but from being driven back into Richmond itself. Lee intended to defend Richmond behind the fortifications of Petersburg, a most important railway junction. But if Grant had at this juncture only had a little luck, the Confederate army would have been driven into the capital. It was, of course, strongly fortified, but it was by no means so strong as Petersburg, and the communications must have been immediately severed.

It is not necessary to explain Grant's perseverance in attacking the Confederates wherever he found them. It is obvious that Lee's army was his true objective, and that the occupation of Richmond could have had no decisive effect while that army still held the field. If that army were thoroughly defeated, the fall of Richmond, and the end of the war, would follow.

I am afraid this is a very imperfect sketch of a very remarkable campaign, but a satisfactory description of these operations would make a fair-sized book. There are, however, a good many points which will bear a little explanation. First of all there is the question of entrenchments.

Defensive tactics, if we are to believe some people, resolve themselves into this:—If you have a point to hold, nothing more is necessary than to take up a position in front of it, to entrench your line till it is as formidable as Plevna, to man it with magazine rifles and machine guns, and to hold on. But I doubt if this is quite enough. I think, on the contrary, that it may be very dangerous, under all circumstances, to select your position long beforehand, and to make sure that the enemy will knock his head against it. Behind the Rapidan, Lee held a very strong entrenched position, covering his line of communications, and covering Richmond. But Grant piles ten days' supply into his 5,000 wagons and walks round the flank of this carefully prepared position. I am particularly anxious not to be misunderstood. I have not the slightest intention, under certain conditions, of denying the very great utility of positions thoroughly prepared and selected long beforehand. Torres Vedras is an instance of their use and value.[6] The lines of Petersburg, occupied by Lee after the Wilderness campaign, are another. But their strength lay in this, that they could not be turned; the line of supply was secure from all attacks, and under such conditions no man in his senses could deny the importance of solidly constructed entrenchments. But there is always the danger—and this is the point on which I am anxious to lay stress—that an army which can manoeuvre like the Federal army by day, and especially by night, an army which can carry large supplies, or which can live on the country, or, above all, which has facilities for changing its base, can often set such entrenchments at defiance. A daring general, like Grant, if he is not tightly bound to one line of supply, will remember Napoleon's maxim, 'shun the position in which the enemy wishes you to attack him, especially that which he has fortified.' Of course it may be said that Lee, in allowing Grant to pass round his flank, and then attacking him in the Wilderness, showed us the best way to deal with such manoeuvres. But this was altogether an exceptional case. Lee relied on the difficulty of the battlefield, on the topography with which he was familiar, and of which his opponents knew next to nothing, and could find out nothing. So greatly was Grant hampered by the lack of roads, that he was unable to reach the open country south of Spotsylvania. Had he possessed greater freedom of manoeuvre, had he not been compelled to move his enormous train by two indifferent roads, it is extremely probable that he would have intervened between Lee and Richmond, and have met him on ground which offered no peculiar advantage, as did the Wilderness, to the Confederates. I am very strongly of opinion that, as modern armies have much practice in manoeuvring, both by day and night, and as their men are trained to long marches, and to movements *en masse*, very careful attention should be directed to the dangers which may arise from the premature selection and occupation of defensive positions. A change of front, especially where large numbers are concerned, if it is to be effected rapidly and in good order, is a most difficult opera-

6. [The lines of Torres Vedras, which spanned the Lisbon peninsula some twenty miles north of that city, were constructed by Wellington in 1810. The French were unable to penetrate these lines and had no choice but to retreat into Spain.]

tion. I may notice here the comparative security in which the Federals manoeuvred by night across the front and round the flanks of the Confederate army.

The country was very close, and reconnoitring parties could not leave the roads in the darkness, but it is impossible not to avoid the conclusion that if, when we occupy a defensive position, we are not desirous of finding the enemy across our flank when the morning dawns, we must use our very best endeavours to find out what is going on under cover of the night. It certainly strikes one as curious that the Confederates, knowing what they did of Grant's predilection for night marches, should have been unable to detect his movement in retreat across the North Anna. This is one lesson, then, which may be deduced from the Wilderness campaign. Because you have formidable earthworks along your front, you are not therefore to consider yourself secure. Another lesson is that the entrenchments which were of the most use in this campaign were those which were constructed on the spot, when the direction of the enemy's attack had become apparent. Those at the Wilderness, Spotsylvania, the North Anna, and Cold Harbor were thrown up when the enemy was actually within striking distance, and yet their value was far greater than those on the Rapidan, or, if Grant's subordinates had been more dashing, than those beyond the James.

For those who care to study the campaign closely, it is worth while noting with what skill Lee's positions were selected. His flanks at Spotsylvania, at the North Anna, and at Cold Harbor, were so secured by streams that it was very difficult indeed for his opponent to manoeuvre without crossing one of these streams, and so dividing his army. It was not only the entrenchments, but the natural features of the ground also on which Lee relied in his defensive tactics. His eye for ground must have been extraordinary. The campaign was fought over a very large area, an area of very close country, with few marked natural features; and yet in the midst of woods, jungle and streams, with very little time at his disposal, he always seems to have selected positions than which none could have been stronger. His eye for ground, then, had much to do with his successful resistance to Grant's overwhelming numbers; and this eye for ground he possessed in common with all generals who are acknowledged as masters of war. Now, with all respect to the text-books, and to ordinary tactical teaching, I am inclined to think that the study of ground is often overlooked, and that by no means sufficient importance is attached to the selection of positions, to the rapid adaptation of hasty entrenchments to the field of battle, to the recognition of 'tactical' points, *i.e.* 'key points'; and to the immense advantages that are to be derived, whether you are defending or attacking, from the proper utilisation of natural features. There are people who tell you that Napoleon's campaigns are ancient history. 'Read the battles of 1870,' they say, 'visit the fields of 1870. There is no use in studying Napoleon's battlefields, because the ranges were so short.' With those good people I altogether disagree. Napoleon, like Lee, made such remarkable use of ground that natural features played a very great part in many of his victories, and no one who visits the scenes of some of these victories can fail to learn a very useful lesson; a lesson of great value to every officer who has any aspirations in the direction of independent command, and this lesson is one in generalship. One of the secrets of Napoleon's extraordinary success will be revealed, and these secrets are well worth the learning, for natural features, as we learn from this very campaign, can still be utilised with great effect, and can be utilised in the very same manner as they were by Napoleon.

Speaking for myself, I may say that I had visited the battle-fields of 1870 very often, and studied them very closely, before I visited any one of Napoleon's fields; but it was not until I went to Jena and Austerlitz that I really grasped what an important part an eye for ground like Napoleon's, or blindness as to ground like his opponents', at both of these battles, may play in Grand Tactics, that is, in the art of generalship. When you look at the position of the Allies at Austerlitz, the position that was captured by one of the finest counter-strokes in history, one of the first things you observe is an insignificant village half-way up the little hill which formed the centre of the position. Napoleon's counter-stroke met with such splendid success because when he saw that village and the hill above it he recognised at once the very great advantage which they would give him if he could seize them. To the ordinary observer they do not appear to be an important point, nor did they seem so to the Allies, who altogether rejected them, or, at all events, took no special precautions for their defence. It seems rather a curious thing to say that we can learn the use of ground from books; but to a certain degree we may learn from the campaigns of the great captains how to utilise the ground; we may learn to recognise its importance; and then proceeding to the ground itself, whether at manoeuvres in command of troops, or in studying positions alone, we can put theory into practice, and gradually acquire that eye for ground without which no man, it is my firm conviction, can ever hope to be a great or even a useful general.

STONEWALL JACKSON'S PLACE IN HISTORY

[EDITOR'S NOTE: The following article was written especially for the second edition of Mary Anna Jackson's biography of her husband.[1] Evidently Major Jed Hotchkiss, who had approached Mrs. Jackson several times on Henderson's behalf, had arranged for the inclusion of this paper in the volume. Henderson sent the manuscript to Hotchkiss with the following confession:

> It has been scribbled off, as fast as I could write, and may not meet with your critical approval; but I don't know that I can better it without longer labours than it is possible to give at present. You will recognize several passages taken straight out of the chapters of my own book which you have so kindly read [in manuscript] and commended.
>
> I have not had the Ms. typ-copied, as time presses, and I don't envy the printer. But I daresay you will be good enough to cast your eye on the proofs, and generally to exercise the functions of an editor. I give you plenary powers to do what you will with both the original and the printer's slips.[2]

After the book was published, Henderson undertook to sell 100 copies in England as a personal favor to Mrs. Jackson.

1. [Mary Anna Jackson, *Memoirs of Stonewall Jackson* (Louisville, 1895), pp. 579–600.]
2. [Henderson to Jed Hotchkiss, October 5, 1895, Hotchkiss Papers, Library of Congress.]

Apparently he could not dispose of these as easily as he had at first anticipated, and he reluctantly confided to Hotchkiss:

> I daresay it will be a slow business, but I hope to get the majority of them disposed of. Lord Wolseley wanted to take the whole lot himself and to hand me over a cheque, but he is as extravagant in the way of kindness as in his admiration for the Confederates, so I absolutely refused to let him do it. He is not well off for his position [Commander-in-Chief of the British Army], and I know that Lady Wolseley would have "combed somebody's hair" if the transaction had been concluded.[3]

The echoes of the Civil War have not yet died away. The survivors of the great conflict still keep its memories green; and we are still privileged to hear, from the lips of those who shared in them, the conversations around the fire of the bivouac, and to learn the opinions of the rank and file on the subjects in which the soldier takes special interest. Foremost among the most absorbing topics of the camp was, undoubtedly, the character of the different generals, whether friend or foe. When one man holds in his hand the lives of thousands, when one word means victory or defeat, the minds of those thousands, even hardened as they may be, must scan with something more than curiosity the individual who rules their fate. The soldier in the ranks tests his commander from two points of view: first, from his achievements, second, from his personality; and than the men who carry the musket there are no shrewder judges. They may be ignorant of the scope of the campaign, of the purpose of the manoeuvres, but they have much to do with their execution. Better than all the historians, better than the higher leaders, they appreciate the difficulties which attend the operations. In their own limbs they have realized the length and labor of the marches; with their own eyes they have seen the strength of the enemy's positions and the numbers that manned them; and their intelligence, rate their military knowledge as you will, is more than sufficient to enable them to recognize a hazardous situation, and to appreciate the exact measure of ability which was brought to bear upon it. They know—and none better—whether the orders of the general were decided and to the point, whether the opportunity was utilized, whether the attack was pressed with resolution, or defence maintained to the ut-

3. [Henderson to Jed Hotchkiss, April 7, 1897, Hotchkiss Papers.]

most limit of endurance. They judge by results. They have seen the enemy driven in panic flight by inferior numbers, his detachments surprised, and his masses outmanoeuvred; and though the victories thus won may have been relatively unimportant, the strength of the opposing forces insignificant, the lists of casualties and prisoners comparatively small, yet the soldiers are not deceived. The world at large recks little of minor engagements, and is much too apt to measure military capacity by the "butcher's bills;" but the instinct of the soldier tells him, and tells him truly, that genius of the highest order may display itself in the defeat of ten thousand men as clearly as in the defeat of ten times that number. When he finds that genius he resigns his individuality, and absolute trust takes the place of speculation. The general in whose soldierly abilities his veterans have implicit confidence, no matter what the scale of his victories, is, without doubt, a leader of men; for that confidence is not easily given, it is only to be won on the perilous edge of many battles, and it is only accorded to consummate skill.

Amongst the echoes of the Civil War there is none of clearer tone than the soldier's estimate of Stonewall Jackson. It never fell to Jackson's lot to lead a great army or to plan the strategy of a great campaign. The operations in the Valley, although far-reaching in their results, were insignificant both in respect of the numbers employed and of the extent of their theatre. Nor was Jackson wholly independent. His was but a secondary role, and throughout the campaign he had to weigh at every turn the instructions or suggestions of his superior officers. His hand was never absolutely free. His authority did not reach beyond certain limits, and his operations were confined to one locality. He was never permitted to "carry the war into Africa." Nor when he joined Lee at Richmond was the restraint removed. In the campaign against Pope, and in the march into Maryland, he was certainly intrusted with tasks which led to a complete severance from the main army; but that severance was merely temporary. He was the most trusted of Lee's lieutenants, but he was only a lieutenant after all. He had never the same liberty of action as Johnston, or Bragg, or Hood; and consequently he had never a real opportunity for revealing the height and breadth of his military genius. What would have been the issue of the war if Jackson had been placed in command of the western armies of the Confederacy, whilst Lee held fast in Virginia, must remain a matter of speculation. One thing is absolutely certain, Lee would never have been able to replace him. As a subordinate he was incomparable. "General Lee," he said, "is a phenomenon, I would follow him blindfold," applying, with his wonderful insight into character, exactly the same words that his own men had come to apply to himself.[4]

It seems that Lee was slower to learn his comrade's worth. Even the Valley campaign, with its long roll of victories, did not at once enlighten him. After Sharpsburg, perhaps with the memory of Jackson's untoward delay on June 27th and again at Frayser's Farm on the 29th still fresh in his memory, he writes: "My opinion of the merits of General Jackson has been greatly enhanced during this expedition. He is true, honest and brave, has an eye single to the good of the service, and spares no exertion to accomplish his object." How different and how significant was his generous cry, not ten months later, when the glories of Chancellorsville were obscured by Jackson's wound: "Could I have directed events, I should have chosen, for the good of the country, to have been disabled in your stead." Yet even after the "Seven Days" to Jackson was committed every enterprise that necessitated a detachment from the army. It was Jackson, with plenary powers, who was sent to check Pope's advance on Gordonsville, to cut his communications at Bristoe Station, to capture Harper's Ferry, to hold the Valley when McClellan advanced after Antietam, and to fall on Hooker's flank at Chancellorsville. The records of the war show abundantly, in the letters which passed between them, how the confidence of the commander-in-chief in his subordinate increased, until, when the news of Hooker's advance on Chancellorsville was reported, Lee could say to one of Jackson's aides-de-camp: "Tell your good general that I am sure he knows what

---

4. [Henderson regarded Lee and Jackson as "twin heroes—you cannot separate them. I am not quite sure," he wrote to Hotchkiss, "that your people really appreciate what a wonder Lee was. I agree with Jackson. He was a 'phenomenon,' and I mean to show it. It is great impudence to say so, but I have studied his campaigns for ten years, and I am of opinion (forgive the conceit) that I know his rank in history and exactly where he ought to be placed better than any one else" (Henderson to Hotchkiss, April 2, 1895, Hotchkiss Papers).]

to do." Nevertheless, the fact that Jackson never held an independent command, and, more than this, his very excellence as a subordinate, have served to diminish his reputation. Swinton, the accomplished historian, speaks of him as follows: "Jackson was essentially an executive officer, and in that sphere he was incomparable, but he was devoid of high mental parts, and destitute of that power of planning a combination and of that calm, broad military intelligence which distinguished General Lee."[5] And Swinton's verdict has been very generally accepted. Because Jackson knew so well how to obey, it is assumed that he was not well fitted for independent command. Because he could carry out orders to the letter, it is implied that he was no master of strategy. Because his will was of iron, and that his purpose, once fixed, never wavered for a moment, we are asked to believed that his mental scope was narrow. Because he was silent in council, not eager in pressing his ideas, and averse to argument, it is implied that his opinions on matters of great moment were hardly worth hearing. Because his simplicity and honesty were so transparent, because he betrayed neither in face nor bearing any unusual power or consciousness of power, it is hastily concluded that he was deficient in the imagination, the breadth and the penetration which are the distinguishing characteristics of great generals.

Yet look at the portraits of Jackson, and ask if the following description is not exactly applicable? "Strength is the most striking attribute of the countenance, displayed alike in the broad forehead, the masculine nose, the firm lips, the heavy jaw and wide chin. The look is grave and stern almost to grimness. There is neither weakness nor failure here. It is the image of the strong fortress, of a strong soul buttressed on conscience and impregnable will." And the face limned here with such power of pen is not the face of a great conqueror or a great ruler, of a Cromwell or a Wellington, but of Dante. The truth is that his quiet demeanor concealed not only a vivid imagination, but an almost romantic enthusiasm for all that was great or pure or true. Nor was Swinton's verdict the verdict of the soldiers of the Civil War. It was not the verdict of Lee—witness his letter

5. [William Swinton, *Campaigns of the Army of the Potomac* ... (New York, 1882), p. 289.]

already quoted. It was not the verdict of the Southern people and it was not the verdict of their foes. It can hardly be questioned, I think, by those familiar with the records of the war, with the ephemeral literature of the time, with the letters and biographies of the actors, that, at the time of his death, Jackson was the leader most trusted by the Confederates and most dreaded by the Federals. Lee was his only rival, but I much doubt whether at the date of Chancellorsville the news of Lee's death would have been received with so much regret in Richmond, or with as much relief at Washington, as was Stonewall Jackson's. Nevertheless, the instinct of the soldiers is hardly sufficient evidence on which to claim for Jackson a place amongst the most famous generals; and for the reason that his theatre of action was limited, it is difficult to assign the rank which he ought to hold. The rank, however, which, had his power been unfettered as that wielded by Lee or Grant, he could in all probability have attained may be inferred from his achievements in a subordinate capacity. Moreover, Jackson was not always inarticulate. To his intimates he confided his own views on the conduct of the war. His active brain, even whilst he was no more than a brigadier, not only anticipated in what manner victories might be best improved, but, maintaining a comprehensive grasp of the whole theatre of events, determined by what means the ultimate triumph of the Confederacy might be secured. These thoughts took shape in definite proposals. And although they were never, I believe, brought to the notice of the supreme authorities, and whilst it is true that it is much simpler to plan than to execute, much easier to advise than to bear responsibility, these proposals at least reveal the breadth of Jackson's mind, his quick perception of the capital object which should have been held in view by the Confederates, and of the weak joint in the Northern harness. To these, as I pass in review the chief events of Jackson's military career, I may be permitted to refer.

The first year of the war gave the Lexington professor but small opportunity. All he was intrusted with he did well, and his tactical ability was cordially recognized by his superiors. Falling Waters, his first essay in arms before the enemy, was an insignificant affair. At Bull Run his brigade displayed a conspicuous part. The quick perception of the advantages of the position on the *eastern* rim of the Henry Hill had much to do with the Confederate victory. Had the brigade been pushed forward to the western rim, it would have been exposed to the full force of the powerful Federal artillery; as it was, placed on the further edge of the plateau, it secured a certain amount of cover, and rendered the attempt of the Northern batteries to establish themselves on the plateau a disastrous failure. Again, although it is hardly alluded to in the official reports, there can be little doubt, at least in the minds of those who have seen the ground and read the narratives, but that the well-timed charge of the Stonewall Brigade was decisive of the issue. Nor can I omit to mention the ready initiative with which the Stonewall Brigade, ordered up to support the troops at the Stone Bridge, was diverted on the march towards the heavy cannonade on the left flank, or the determined bearing which inspired his defeated colleagues with renewed confidence. If two opinions exist as to the effect of Jackson's charge there can be no question that, but for his ready intervention and skilful choice of a position, the key-point of the battle-field would have been lost to the Confederates. Why the Southern generals did not follow up their success is a question round which controversy has raged for many a year. The disorganization of the victorious volunteers, the difficulties of a direct attack on Washington, deficiencies of supply and transport, have all been pressed into service as excuses. "Give me ten thousand fresh troops," said Jackson, as the surgeon dressed his wounds after the battle, "and I would be in Washington to-morrow." Within twenty-four hours the ten thousand had arrived. There were supplies, too, along the railway in the rear, and if means for their distribution and carriage were wanting, the counties adjoining the Potomac were rich and fertile. It was not a long supply train that was wanting, not a trained staff, nor well-disciplined battalions, but a general who grasped the full meaning of victory, who understood how a defeated army, more especially one of raw troops, yields at a touch, who knew "that war must support war," and who, above all, realized the necessity of giving the North no leisure to develop her immense resources. That Jackson was such a general may be inferred from his after career.

His daring judgment never failed to discern the strategical requirements of a situation, and no obstacle ever deterred him from aiming at the true objective. Whilst in camp after Bull Run he said nothing. Afterwards, to his intimates, he condemned the inaction of his superiors with unusual warmth and emphasis. Of the accuracy of his insight the letters of General McClellan, hurried from West Virginia to command at Washington, are the best evidence. On July 26th, the fifth day after the battle, McClellan "found no preparations for defence. All was chaos. . . . There was nothing to prevent a small force of cavalry riding into the city. . . . If the Secessionists attached any value to Washington, they committed their greatest error in not following up the victory of Bull Run."

Jackson's removal in the late autumn to the Shenandoah Valley was unmarked for some months by any striking incident. The Romney expedition did little more than frighten the Federals and reveal the defects of the raw Confederate soldiers. But during this time Jackson's brain was alive to more momentous questions than the retention of a few counties. The importance of the northwestern districts of Virginia as a recruiting ground, the necessity of an active offensive on the part of the Confederate government, of anticipating the vast preparations of the North, and of bringing the horrors of war home to the citizens of the United States—such questions constantly occupied his mind. But the young brigadier had no voice in the councils of the South. At the end of February began that series of operations which are combined under the title of "The Valley Campaign;" and this campaign, on which Jackson's fame as a master of strategy chiefly rests, was the most brilliant exhibition of generalship throughout the war. As regards this campaign, however, a certain amount of misconception exists. Its success is not to be attributed wholly and solely to Jackson. It was due to Johnston that Jackson was retained in the Valley when McClellan moved to the Peninsula, and his, too, was the fundamental idea of the campaign, that the Federals should be retained in the Valley. It was Lee who at the end of April urged Jackson to strike a blow at Banks, reinforcing the army of the Valley with Ewell's division for that purpose. It was Lee who saw the diversion that might be effected if Jackson threatened

Washington, and it was Lee who exactly at the right moment ordered the Valley troops to Richmond. But it was none the less true that Jackson realized the situation just as clearly as Lee or Johnston. He saw from the very first the weak point in McClellan's plan of campaign, and the probable effect of a threat against Washington. When Lee urged him to strike Banks at Harrisonburg he was already looking for an opportunity. When Ewell arrived it was in response to his own request for reinforcements, and it may be remembered that Lee made no suggestion whatever as to the manner in which his ideas were to be carried out. Everything was left to Jackson. The swift manoeuvres, which surprised in succession his various enemies, emanated from him alone. It was his brain that conceived the march by way of Mechum River Station to McDowell, the march that surprised Fremont and bewildered Banks. It was his brain that conceived the sudden transfer of the Valley army from one side of Massanutton to the other, the march that surprised Kenly and drove Banks in confusion across the Potomac. It was his brain that worked out the design of threatening Washington with such extraordinary results. To him, and to him only, was due the double victory of Cross Keys and Port Republic. If Lee's strategy was brilliant, that displayed by Jackson on the minor theatre of war was no less masterly.

In March, 1862, 200,000 Federals were prepared to invade Virginia. McClellan, before McDowell was withheld, reckoned on placing 150,000 men at West Point; there were 20,000 in West Virginia, and Banks had 30,000 in the Valley. At no time did the army opposed to them exceed 80,000, yet at the end of June where are the "big battalions?" One hundred thousand men are retreating to their ships on the James. But where are the rest? Where are the 40,000 men that should have reinforced McClellan? How comes it that the columns of Fremont and Banks are no farther south than they were in March; that the Shenandoah Valley still pours its produce into Richmond; that McDowell has not yet crossed the Rappahannock? What mysterious power has compelled Lincoln to retain a force larger than the whole Confederate army "to protect the national capital from danger?" Let Kernstown and McDowell, Winchester, Cross Keys and Port Republic speak. The brains of two great leaders had done more for the Confederacy than 200,000 soldiers had done for the Union. Without quitting his desk, and leaving the execution of his plans to Jackson, Lee had relieved Richmond of 100,000 Federals. Jackson, with a force of never more than 17,000, had neutralized and demoralized this enormous force, and, finally joining the main army, had aided Lee to drive the remaining 100,000 away from Richmond.

Nor was this result due to hard fighting alone. The Valley campaign lost the Federals no more than seven thousand men, and, with the exception of Cross Keys, the battles were well contested. It was not due to inferior leading on the battle-field, for at Kernstown, McDowell, Winchester and Port Republic the Federal troops were undeniably well handled. Nor was it due to the want of will on the part of the Northern government. It was simply due to the splendid strategy of Lee and Jackson. Jackson's long and rapid marches were doubtless a factor of much importance; but more important still was the skill that enabled him to effect surprise after surprise, to use the mountains to screen his movements, and on every single battle-field, except Kernstown and Cross Keys, despite the overwhelming superiority of his opponent on the whole theatre, to concentrate a force greater than that immediately opposed to him. "As a strategist," says Dabney,[6] "the first Napoleon was undoubtedly Jackson's model. He had studied his campaigns diligently, and he was accustomed to remark with enthusiasm on the evidences of his genius." "Napoleon," he said, "was the first to show what an army could be made to accomplish. He had shown what was the value of time as a strategic combination, and that good troops, if well cared for, could be made to march twenty-five miles daily and win battles besides." And he had remarked more than this. "We must make this campaign," he said at the beginning of 1863, "an exceedingly active one. Only thus can a weaker country cope with a stronger; it must make up in activity what it lacks in strength. A defensive campaign can only be made successful by taking the aggressive at the proper time.

6. [Robert Lewis Dabney, a Presbyterian clergyman who served on Jackson's staff, provided Henderson with "exhaustive memoranda" on many episodes of Jackson's career (Henderson, *Stonewall Jackson*, I, xix). His *Life of . . . Jackson* (1864) was one of the first biographies written about "Old Jack."]

Napoleon never waited for his adversary to become fully prepared, but struck him the first blow." It would be perhaps difficult in the writings of Napoleon himself to find a passage which embodies his conception of war in terms as definite as these, but no words could convey it more clearly. It is such strategy as this that "gains the aid of States and makes men heroes." Napoleon did not discover it. Every single general who deserves to be entitled great has used it. It was on the lines here laid down that Lee and Jackson acted. Lee, in compelling the Federals to keep their columns separated, manoeuvred with a skill which has seldom been surpassed. Jackson, falling as it were from the skies into the midst of his astonished foes, struck right and left with extraordinary swiftness, and with seventeen thousand men paralyzed, practically speaking, the whole Federal host. It is when regarded in connection with the operations of the main armies that the Valley campaign stands out in its true colors; but at the same time, as an isolated incident, it is a campaign than which few can show more extraordinary results. It has been compared, and not inaptly, with the Italian campaign of 1796; in some of its features it resembles that of 1814; and in the secrecy of movement, celerity of march, the skilful use of topographical features, in the concentration of inferior force at the critical point, it bears strong traces of the Napoleonic methods. Above all, it reveals a most perfect appreciation of the best means of dealing with superior numbers. The emperor could hardly have applied his own principles with more decisive effect.

Moreover, like that of 1796, the Valley campaign was carried through by an officer who had but scant experience of command. Like Napoleon when he dashed through the passes of the Apennines, driving Austrian and Sardinian before him, Jackson in 1862 had served no long apprenticeship to war, and yet his first important enterprise, involving most delicate questions of strategy and supply, was carried to a successful conclusion in the face of an enemy who at one time was trebly superior, and takes rank as a masterpiece of leadership. It is possible that Jackson, in one characteristic, even excelled Napoleon. With all his daring he was pre-eminently cautious. He was neither intoxicated by victory nor carried away by the *gaudia certami-*

nis. His self-restraint was as strong as Wellington's. Like the great Englishman, he knew as well when to decline a battle as when to fight one; he was never inveigled into a useless conflict, and his triumphs were never barren. The whole Valley campaign—from Kernstown to Port Republic—cost the Confederacy no more than twenty-five hundred men; and this economy of life was due as much to Jackson's prudence as to his skilful strategy. He never forgot that his was but a secondary role; that the decisive act of the campaign must be played before Richmond, and that every available musket would be needed to overwhelm McClellan. It is easy to imagine how his patience must have been tried when Fremont, after Port Republic, fell back on Harrisonburg; how every impulse of his being must have urged instant pursuit; how every soldierly instinct must have told him that the prey was before him and that it needed but a few swift marches to crown the campaign by a victory more complete than any he had already won.

The Valley campaign may be said to have been Jackson's only opportunity for showing his strategical ability. In the movements (July 19th to August 14th) against Pope, culminating in the battle of Cedar Run, although he completely achieved his object, the situation demanded no pre-eminent abilities. The Federal commander, in pushing Banks forward without support, committed a mistake, and Jackson, with his usual promptness, took swiftest advantage of it. The second phase of the campaign, however, gave a more brilliant opening. Thrust with his single corps astride the enemy's communications, with his back to the Bull Run Mountains, the remainder of the Confederate army still beyond the passes of that outlying range, and Pope's masses rapidly converging on his isolated troops, he had to face a situation that few would have faced unmoved. The manoeuvres by which he baffled his adversaries, slipped from between their fingers, and regained his connection with Lee at exactly the right moment, were even more skilful than those in which he escaped the converging columns of Fremont and McDowell at the end of May. Had the worst come to the worst he could always have retired through Aldie Gap; but Lee's object—the immediate overthrow of Pope before he could be reinforced by McClellan—forbade retreat, and Jackson's brains and energy were equal to the task. A month later Lee imposed on him the capture of Harper's Ferry. It was carried out, as were all of Jackson's operations, in a manner which defies criticism, and throughout, the requirements of the general situation, the danger which menaced the main army, were foremost in his mind. With the fall of Harper's Ferry the tale of Jackson's detached enterprises came to an end.

This is hardly the place to discuss his views on the military policy of the Confederate government. He was an ardent and consistent advocate of invasion, and I have already quoted his conviction as to the only sound course which can be pursued by the weaker side. On this point opinions will probably differ, but it may be said that it is a course which has the sanction of many precedents, and has been the invariable practice of the great masters of war. Nor can I do more than refer to the methods by which Jackson proposed to bring the North to its knees. They are fully explained in Mrs. Jackson's pages, and to examine their merits and to weigh their probable chances of success would be to write a treatise on the war.

So far I have confined myself to Jackson's conception and application of strategic principles. That both conception and application could hardly have been improved upon is my firm conviction. It is difficult to point out even the shadow of a mistake. Nor was Jackson the tactician inferior to Jackson the strategist. Space forbids me examining the salient features of his many battles; but from Kernstown to Chancellorsville the same characteristics almost invariably reappear. Concentration of force against the enemy's weakest point, the employment at that point of every available man and gun, a close combination of the three arms, infantry, cavalry and artillery, relentless energy in attack, constant counterstroke on the defensive, were the leading principles on which he acted; and here again he was Napoleonic to the core. It has been said that the leaders of the Army of the Potomac, as Lincoln's native shrewdness detected, never "put in all their troops." Even Grant, in the campaign of 1864, failed, except at Cold Harbor, in this respect, and at Cold Harbor the troops were not put in at the enemy's weak point. Here Jackson never blundered, and we may compare the strength of the three lines which crushed Hooker's left at Chancellorsville[7] with the comparative weakness of the assault at Gettysburg; and yet the Federal army at Chancellorsville was stronger and the Confederate weaker than on July 3d. It is true that Jackson was not invariably tactically successful. He was beaten at Kernstown, although that action was a strategic success; his advanced guard was roughly handled at McDowell; Port Republic might well have been a less costly victory, and at Frayser's Farm his delay was disastrous.

To my mind, however, the action with Gibbon at Gainesville,[8] although the troops behaved magnificently, was the only occasion on which Jackson showed less than his wonted skill. His delay at Frayser's Farm is explained by his letter to Mrs. Jackson. . . . Constant rain and unhealthy bivouac had brought on an attack of fever;[9] but at Gainesville the tactical disposition of the Confederate forces was not such as we should have looked for. It was purely "a hammer and tongs" fight, carried through with extraordinary gallantry by the men, but with no manoeuvring whatever on the part of the Confederate general.

Napoleon, however, wrote, "I have made so many mistakes that I have learned to blush for them," and the specks on Jackson's fame as a tactician are not only few and far between, but may generally be attributed to the shortcomings of his subordi-

7. I have not entered into the vexed question of whether Lee or Jackson designed this movement, and I am convinced in my own mind that both saw the weak point in the Federal dispositions, just as they had both seen the weak point in 1862. [For a discussion of who devised the flank movement at Chancellorsville, see Douglas Southall Freeman, *R. E. Lee* (New York, 1935), II, Appendix V; Sir Frederick Maurice, *Robert E. Lee the Soldier* (Boston and New York, 1925), pp. 184–86. Both arrive at substantially the same conclusion: the basic plan was Lee's, the admirable execution was Jackson's.]

8. [Generally known as the battle of Groveton, fought August 28 and 29, 1862.]

9. [See Mary Anna Jackson, *Memoirs of Stonewall Jackson*, p. 302.] Since writing this I have studied the Battles of the Seven Days, and the difficulties and circumstances in Jackson's report in greater detail, and have come to the conclusion that Jackson could not have joined in the battle, and that he was perfectly right in not attempting to throw his infantry over White Oak Swamp in the face of Franklin's troops. [This footnote was added at the last minute by Hotchkiss, at Henderson's urgent request (Henderson to Jed Hotchkiss, December 26, 1895, Hotchkiss Papers). Correspondence with the Rev. Robert Lewis Dabney, who had served on Jackson's staff, also seemed to justify Jackson's slowness at Gaines' Mill, the second of the Seven Days battles. "His account," Henderson confided to Hotchkiss, ". . . almost lifts a weight from my mind. I have so profound a faith in Jackson's genius that I cannot convince myself that he made the slightest mistake, and I am delighted to find that in this instance my instinct was right. I feel more that it was some one else who had blundered" (Henderson to Jed Hotchkiss, April 5, 1896, Hotchkiss Papers).]

nates or to the unavoidable accidents of war. One point as regards Jackson's tactical skill has hardly received sufficient attention. Although his whole knowledge of cavalry was purely theoretical, he handled his squadrons with an ability which no other general up to the date of his death had yet displayed. I am not alluding merely to the well-timed charge which captured Kenly's retreating infantry after the engagement at Front Royal, although that in itself was a brilliant piece of leadership, but to the use made of the cavalry in the Valley campaign. It is true that Stuart had already done good work in 1861, but as a general commanding a force of all arms Jackson was the first to draw the full benefit from his cavalry.

"The manner," says Lord Wolseley, "in which he mystified his enemies is a masterpiece." It was not, however, his secrecy regarding his plans on which he principally relied to keep his enemy in the dark. Ashby's squadrons were the instrument. Not only was a screen established which perfectly concealed the movements of the Valley army, but constant demonstrations at far distant points confused and bewildered the Federal commanders. In his employment of cavalry Jackson was in advance of his age. Such tactics had not been seen since the days of Napoleon. The Confederate horsemen in the Valley were far better handled than those of France or Austria in 1859, of Prussia or Austria in 1866, of France in 1870, of the Allies or the Russians in the Crimea. In Europe the teachings of Napoleon had been forgotten. The great cloud of horsemen which veiled the marches of the Grand Army had vanished from memory; the great importance ascribed by the emperor to procuring early information of his enemy and hiding his own movements had been overlooked; and it was left to an American soldier to revive his methods. Nor was Jackson led away by the specious advantages of the so-called "raids." In hardly a single instance did such expeditions inflict more than temporary discomfort on the enemy, and more than once an army was left stranded and was led into false manoeuvres for want of the information which the cavalry would have supplied.

Hooker at Chancellorsville, Lee at Gettysburg, Grant at Spotsylvania, owed defeat, in great measure, to the absence of their mounted troops. In the Valley, on the contrary, success was made possible because the cavalry was kept to its legitimate duty, that is, to procure information, to screen all movements, and to take part in battle at the decisive moment.

Jackson was certainly fortunate when Ashby came under his command. That dashing captain of free lances was a most valuable colleague. It was much to have a cavalry leader who could not only fight and reconnoitre, but who had capacity enough to divine the enemy's intentions. But the ideas that governed the employment of cavalry were Jackson's alone. He it was who, at the end of May, placed the squadrons across Fremont's road from Wardensburg, who ordered the demonstrations against Banks and those which caused Fremont to retreat after Port Republic. More admirable still was the quickness with which he recognized the use of cavalry that could fight dismounted. From the Potomac to Port Republic his horsemen covered his retreat, lining every stream and the borders of every wood, holding on to every crest of rising ground, checking the pursuers with their fire, compelling them to deploy, and then withdrawing rapidly to the next position. Day after day was Fremont's advanced guard held at bay, his columns delayed, and his generals irritated by their slippery foe. Meanwhile the Confederate infantry, falling back at their leisure, were relieved of all annoyance. And if the cavalry were suddenly driven in, support was invariably at hand, and a compact brigade of infantry, supported by artillery, quickly sent the pursuers to the rightabout. The retreat of the Valley army was managed with the same skill as its advance, and the rearguard tactics of the campaign are no less remarkable than those of the attack.

I have said nothing about Jackson's marches, and, as a matter of fact, while he managed to get more out of his men than any other commander of his time his marches can hardly be classed as extraordinary. They certainly do not exceed those made elsewhere; and if it be asserted that the Virginian roads are bad, they could hardly have been more infamous than those travelled by both the French and English troops in Spain and Portugal; and yet the marches in the Peninsula, on very many occasions, were longer and more rapid than those of "the foot cavalry."

When Jackson fell at Chancellorsville, his military career had only just begun, and the question, what place he takes in history, is hardly so pertinent as the question, what place he could have taken had he been spared. So far as his opportunities had permitted, he had shown himself in no way inferior to the greatest generals of the century, to Wellington, to Napoleon or to Lee. That Jackson was equal to the highest demands of strategy his deeds and conceptions show; that he was equal to the task of handling a large army on the field of battle must be left to conjecture; but throughout the whole of his soldier's life he was never intrusted with any detached mission which he failed to execute with complete success. No general made fewer mistakes. No general so persistently outwitted his opponents. No general better understood the use of ground or the value of time. No general was more highly endowed with courage, both physical and moral, and none ever secured to a greater degree the trust and affection of his troops. And yet, so upright was his life, so profound his faith, so exquisite his tenderness, that Jackson's many victories are almost his least claim to be ranked amongst the world's true heroes.

THE HENDERSON LEGACY

In August, 1898, Henderson completed his major work, *Stonewall Jackson and the American Civil War*. For over a decade he had worked on this labor of love, studying the *Official Records*, corresponding with numerous former Confederates, and scrutinizing every source he could lay his hands on. Major Jed Hotchkiss, Jackson's topographical engineer, had proved indispensable in supplying maps and unpublished reports and especially in answering Henderson's innumerable queries. Did Jackson, in planning his maneuvers, depend upon his maps or his memory of the terrain? What sort of books did he have in his library? ("It will be a very interesting point to all thinking soldiers over here.") Had he studied the campaigns of Napoleon? How did he look in combat? "What is the truth of the delay on June 25 (night) and 26th?"[1] Hotchkiss faithfully answered these and many similar questions, and he put Henderson in touch with other former Confederates who could add touches to the portrait of their esteemed leader. "Never was a biographer better served than by these men."[2] From these varied

1. Henderson to Jed Hotchkiss, April 24, 1892, April 11, 1895, April 28, 1895, June 27, 1895, Hotchkiss Papers, Library of Congress.
2. Douglas Southall Freeman, *The South to Posterity* (New York, 1951), p. 161. For specific information furnished by Hotchkiss see *Stonewall Jackson*, I, 219, 416–17; II,

sources Henderson was able to capture the spirit of Stonewall and his men as no foreign writer—and few American writers—have succeeded in doing before or since. Stonewall Jackson came to life, and his campaigns acquired a new significance.

Henderson's biography of Jackson is a classic in several respects. A prominent reviewer correctly described it as a military biography, an authentic campaign history, and a general treatise on the art of war, all in one![3] "As a biography it is a model, and as such it may be read with pleasure by those for whom the details of the campaign may not have any great interest."[4] As a military history it has several weaknesses. Henderson dodged controversial questions at times, and he so admired Jackson that he was reluctant to admit that even Jackson made occasional errors.[5] But these faults were more than offset by Henderson's grasp of the subject and his insight into the military conditions of the Civil War.[6] His treatment of the Shenandoah Valley campaign of 1862 has served as a model for later writers.

Probably few people realize that *Stonewall Jackson* was intended to serve as a treatise on the art of war, but its pages contain Henderson's thoughts on virtually every phase of military activity. Henderson believed that Hamley's *Operations of War* had grievous shortcomings as a military text. Hamley, he

---

431–32, 436. Former Confederate staff officers J. Hunter Maguire, J. P. Smith, and Generals G. W. Smith, Fitzhugh Lee, Stephen D. Lee, and N. G. Harris, as well as lesser officers too numerous to mention here, likewise contributed to Henderson's research.

3. Lt. Gen. Sir Henry Brackenbury, "Stonewall Jackson," *Blackwood's Edinburgh Magazine*, CLXIV (December, 1898), 722 (cited hereafter as Brackenbury, "Stonewall Jackson").

4. Roberts, "Memoir" (see chap. i, n. 2, above), p. xxxii.

5. With reference to Jackson's tardy arrival at the battle of Beaver Dam Creek, Henderson inquired of Hotchkiss: "What is the truth of the delay. . . . Dabney says some subordinates blundered. This is not enough. We want the truth, and we want to exonerate Jackson" (Henderson to Hotchkiss, April 11, 1895, Hotchkiss Papers). "The truth," Henderson ultimately decided, "is that the arrangements made by the Confederate Headquarter staff were most inadequate" (*Stonewall Jackson*, II, 20). Freeman mentions another of Henderson's weaknesses, an "uncertainty regarding some terrain," but he adds that, with the evidence available at the time, Henderson could hardly be expected to have covered Jackson's operations more thoroughly (*The South to Posterity*, pp. 160–64).

6. "Better perhaps than any foreign critic of the war . . . Henderson understood the subtle military values that gave the North a crushing superiority over the South" (*ibid.*, pp. 164–65).

felt, had "deliberately omitted all reference to the spirit of war, to moral influences, to the effect of rapidity, surprise and secrecy"—those vital intangibles so essential to good leadership and, incidentally, so characteristic of Jackson. Speaking before the Royal United Service Institution in 1894, Henderson explained his views on the need for a new textbook in strategy.

> The methods by which the great generals bound victory to their colours are scarcely mentioned in the tactical text-books; and in Hamley's 'Operations of War' the predominating influence of moral forces is alluded to only in a single paragraph. In short, the higher art of generalship, that section of military science to which formations, fire, and fortifications are subordinate ... has neither manual nor text-book.[7]

There can be no doubt that Henderson, in writing his biography of the great Confederate leader, had in mind also the need for filling this gap in military literature.

Instead of outlining campaigns and expounding principles, Henderson wrote from the eye level of Jackson, describing each situation as he thought Jackson himself would have viewed it and focusing his attention upon the methods and psychological reactions of the commander. This technique enabled Henderson to inject his own philosophy of war so skilfully into the narrative that on occasions where the facts were not known—or did not appear to reflect the usual credit on the Confederate general—Jackson's actions are explained and justified by Henderson's own strategical concepts.[8] While this practice occasionally led to minor distortion of fact, it did not necessarily detract from the value of the book as a military study. On the contrary, as Captain B. H. Liddell Hart has suggested, it probably even enriched it as a military treatise, since the genius of Jackson was supplemented by the theories of a profound student of war.[9]

7. *The Science of War*, p. 169; A. R. Godwin-Austin, *The Staff and the Staff College* (London, 1927), p. 114. See also Henderson, "Strategy and Its Teaching," *Journal of the Royal United Service Institution*, XLII (July, 1898), 767.

8. A case in point is Henderson's account (*Stonewall Jackson*, I, 328 ff.) of Jackson's decision to march to Middletown on the Valley Pike immediately following the capture of Front Royal, Virginia, on the night of May 24, 1862. According to Freeman, this situation even today cannot be given the clear treatment that Henderson provided when he grafted his own thoughts to Jackson's action and thus probably credited Jackson with undue skill in this particular operation (statement made by Douglas Southall Freeman to the writer, March 28, 1950, in Richmond, Virginia). For a similar example of this transplanted reasoning, see p. 228 and n., above.

9. *The British Way in Warfare* (London, 1932), p. 76.

Stonewall Jackson was greeted with enthusiastic acclaim both in England and in the United States. At first Henderson professed disappointment with its reception in the army,[10] but soon it was indorsed by such prominent soldiers as Lord Wolseley and Field Marshal Earl Roberts (who later claimed that the book had provided him with inspiration and guidance in his campaign against the Boers[11]). In the United States, reviewers described it as "the most impartial work yet published on the Civil War," "an art-study in war," and predicted that the book would live "as long as the language is spoken." A former Confederate staff officer who had been close to Jackson commented that Henderson "evidently has no bias in favor of the cause for which Jackson died"; Mrs. Jackson was said to have been shocked by the comparison between her husband and Lincoln, but otherwise she thought very highly of the book. The general impression created by its publication is illustrated by the comments of Sir Henry Brackenbury, who wrote, "As an old Professor of Military History, I uncover my head to the author, and tender him my grateful thanks."[12]

In 1899 war broke out in South Africa. After the British had suffered a series of setbacks, Lord Roberts was called to assume command. He had previously heard Henderson lecture before the Dublin Military Society, and by coincidence he had just finished reading *Stonewall Jackson*. Convinced that Henderson "would be able to turn his knowledge to practical account," Roberts, in January, 1900, appointed him to his newly formed

10. Soon after publication of *Stonewall Jackson*, Henderson complained to Sir Henry Brackenbury: "Besides yourself, only one other General, or even Colonel, on the active list has said a word about the book to me; and so far from its having attracted any notice, I find that very few of the senior officers even know that it is in existence. We are certainly not a literary army, and the unfortunate soldier with a turn for writing history does not get much encouragement from the service. The Volunteers, however, are noble creatures: they actually buy military books, and spend their money freely in educating themselves, so there is hard cash to be made out of writing, and that is a consolation" (Sir Henry Brackenbury, *Some Memories of My Spare Time* [Edinburgh and London, 1909], p. 86).

11. Roberts, "Memoir," pp. xxxiv–xxxv.

12. See S. S. P. Patteson, "Henderson's *Stonewall Jackson*," *Sewanee Review Quarterly*, VII (1899), 88; Henry Kyd Douglas, "Stonewall Jackson and the American Civil War," *American Historical Review*, IV (January, 1899), 372. Mrs. Jackson's reaction is mentioned in a letter from Jed Hotchkiss to Henderson, December 1, 1898, Hotchkiss Papers. See also Brackenbury, "Stonewall Jackson," p. 722.

staff as Director of Intelligence. On the long sea voyage out, the two spent hours together walking up and down the decks of the "Dunottar Castle," discussing strategy and its applications to the coming campaign. During the days of preparation at Cape Town, Henderson's "fertile suggestions and sober criticisms . . . played no small part in confirming the native intuition and strengthening the resolution of his chief."[13] In South Africa Henderson performed useful service in obtaining maps of the theater of operations, and, doubtless guided by Jackson's dictum "always mystify and mislead" (a phrase that has become idiomatic in English military parlance), he even took pains to plant misleading newspaper articles to camouflage the British plan of campaign. Lord Roberts has recorded that during the campaign former students at the Staff College would file into Henderson's tent at odd hours, "eager to discuss those actual problems which they had so often studied in theory, glad of the chance given them of referring their doubts and difficulties" to their esteemed teacher.[14]

But ill health soon caused Henderson to give up active campaigning and return to England, where he was assigned the task of writing the official history of the war. In the fall of 1901 he again visited South Africa, this time to study the battlefields and gather material for his book. He soon suffered a relapse, and although he continued with his research for another year he did not recover. Never in robust health, Henderson died in 1903 in Egypt, where he had been sent to avoid the rigors of another English winter. He was not quite fifty.

"The influence of such a man must bear good fruit." Lord Roberts expressed his hope that Henderson's writings would have as much influence on British military policy as the books of Captain A. T. Mahan, the famed American naval historian, had had in naval matters.[15] In the obituary that appeared in the *Times*, Henderson's influence in the British army was likened to that of Von Moltke in Germany.[16] While this statement is doubtless an exaggeration, Henderson did, in fact, have a wide

13. L. S. Amery, *My Political Life* (London, 1953), I, 126. Amery was chief war correspondent for the *Times* during the Boer War.
14. Roberts, "Memoir," pp. xxxvi–xxxvii.
15. *Ibid.*, p. xxxiii.
16. *Times*, March 7, 1903.

following, and his views on military subjects were well received. Through his lectures and numerous publications, he reached a large audience, and for years he served as the regular military correspondent for the *Times* in covering foreign military maneuvers. In 1901, Lord Roberts ordered him to revise and bring the old *Infantry Drill Book* up to date. Henderson died before completing this task, but the chapters he left behind were found to be so meaty in doctrine common to all the arms that the committee designated to complete the work issued it under a new title, *Combined Training* (1902). The "forerunner of a new conception of military textbooks," *Combined Training* represented a definite break with the past. Lord Roberts claimed that "all works and regulations which had hitherto dealt with the subject contained in this manual either have been, or shortly will be revised," and gave instructions that it was to be regarded as "authoritative on every subject with which it deals."[17]

From the nature of his writings, however, it seems probable that Henderson left his deepest mark not so much upon tactical and material reforms as upon military thought in general. His writings, like his lectures, were inspirational. *Stonewall Jackson*, "an admirable exposition of the generalship of small semi-professional armies," was used widely as a military text. Lord Wolseley stated that he knew of no book which would add more "to the soldier's knowledge of strategy and the art of war," and General Fuller has testified that "Its influence was enormous."[18]

Henderson has not become identified with any specific school of military theory, and for good reasons. He died before his thoughts on war had crystallized: *The Science of War* is merely a collection of representative essays and not a carefully prepared synthesis of his views, and in all his writings he was interested primarily in providing "a clear insight into the innumerable problems connected with the organisation and the command of an armed force."[19] It must be remembered, too, that while

17. Dunlop, *The Development of the British Army*, pp. 225–26, 291. *Combined Training*, in 1905, became Part I of *Field Service Regulations*, which was superseded by a new manual in 1909. Sir James Edward Edmonds (comp.), *Military Operations France and Belgium, 1914* (London, 1922), I, 9.

18. Wolseley, "Introduction," *Stonewall Jackson*, I, xvi; Fuller, *The Army in My Time*, p. 122.

19. *The Science of War*, p. 395.

Henderson envisaged a war fought by a British army "far larger" than that which Wellington had commanded, he was writing for an army designed primarily to police the empire and protect the country against the old bogy of invasion. In Germany and France, the military leaders could anticipate every detail of mobilization beforehand and shape their strategy accordingly. But in England it was not so simple:

> It is as useless to anticipate in what quarter of the globe our troops may be next employed as to guess at the tactics, the armament, and even the colour ... of our next enemy. Each new expedition demands special equipment special methods of supply and special tactical devices, and sometimes special armament.... Except for the defence of the United Kingdom and of India, much remains to be provided when the Cabinet declares that war is imminent.[20]

In his writings Henderson was as flexible as the military situation confronting Britain. Strategical principles were to be obeyed "rather in the spirit than in the letter." According to Henderson, the successful strategist is a man like Jackson, who knows "exactly how far he can go in disregarding or in modifying" the so-called principles of war, and is at the same time "ingenious enough to bring those into adjustment which are apparently irreconcilable."[21] Henderson's students faced a more rigid situation. The shift in British foreign policy signified by the Anglo-French Entente in 1904 meant not only the abandonment of a policy of isolation but, ultimately, commitment to France. It became increasingly clear that the army these men had to raise, train, and equip was destined for use on the Continent. Henderson had indicated where answers to some of the questions might be found; he did not, however, devise any specific formula for victory.

Henderson also left his mark on military education. As one of the team that Lord Wolseley sent to the Staff College in the 1890's to make the British officer a more serious student of his profession, Henderson made a major contribution in elevating the reputation of that institution. Previously the courses there had often been dull and formal. Sir George Aston, who attended the Staff College in 1889, recalls that one of the best students

20. [Henderson], "The War in South Africa," *Edinburgh Review*, CXCI (January, 1900), 251–52.
21. *The Science of War*, p. 42.

in his class, a Royal Engineer "who had earned great fame by his surveys of large areas on the North West Frontier of India," had had one of his maps returned by the instructor with the comment, "You should practise gravel-pits."[22] But with the arrival of the new regime such pedantry was discouraged. "We want officers to absorb, not to cram," insisted Colonel H. T. J. Hildyard, who became Commandant in 1893, and Henderson's teaching was in line with this policy. He broadened the course in military history to include the Civil War campaigns (previously the wars of 1866 and 1870–71 had been the chief objects of study), and, perhaps remembering his days as a history student at Oxford, he gave each student personal attention such as he might receive in a university seminar.[23]

Henderson must have been an unusual teacher, and there is abundant evidence to indicate that he was an effective one. By "the charm of his personality and the inspiration of his teaching" he exercised an influence that was "almost unique" in the history of the Staff College. He was intrusted with some of the best minds in the army, and many of his students later reached positions of high command during World War I. These men, wrote one of Henderson's most prominent pupils, "would readily admit that such successes as attended their leadership were largely due to the sound instruction and inspiring counsel which they received from their old tutor some twenty years or so before." The *Times* had sounded the right note indeed when it had predicted, in 1903, that Henderson's influence would be felt "in the next great war, if that should take place when those who have passed through the Staff College in the nineties are in positions of command."[24]

Henderson's writings, particularly *Stonewall Jackson*, influenced both the quantity and the quality of the literature on the Civil War published in England during the following decade. Most of the books and articles that appeared were written to assist hopeful young officers to master the intricacies of Jack-

22. Major General Sir George Aston, *Memories of a Marine* (London, 1925), pp. 100–101.
23. Godwin-Austin, *The Staff and the Staff College*, pp. 231–33.
24. Sir William Robertson, *From Private to Field Marshal* (Boston, 1921), p. 83; Sir Frederick Maurice, *The Life of General Lord Rawlinson of Trent* (London, 1928), pp. 26, 84; *Times*, March 7, 1903.

son's campaigns in preparation for the Officer Promotion Examinations.[25] The military history section of these examinations generally included one or more of the Civil War campaigns, and each year that a Civil War topic was selected new studies appeared about the campaign in question. Both the examinations themselves and the accompanying cram books leaned heavily upon the facts and opinions recorded in Henderson's biography of Jackson. In November, 1904, for example, the questions for the promotion examinations were based upon Henderson's treatment of the Virginia campaigns, with special emphasis upon Jackson's operations in the Shenandoah Valley. A similar examination for officers of the Auxiliary forces covered the operations leading up to and including the battle of Fredericksburg, Henderson's books being considered the chief text.[26] The questions for 1908–9 concerned Grant's campaign in Virginia in 1864,[27] but the following year officer candidates were again

25. The Promotion Examinations were instituted in 1850, when it was decreed that ensigns and lieutenants should undergo an examination before being promoted to a higher rank (Sir John W. Fortescue, *A History of the British Army* [London, 1899–1930], XIII, 20–21). In 1871, a system of garrison classes to prepare officers for these examinations was developed, and years later, when these classes were abolished, young officers had to resort to the "crammers" (Colonel Willoughby Verner, *The Military Life of H. R. H. George, Duke of Cambridge* [London, 1905], I, 143–45). Most of the Civil War campaign studies written before 1914 were designed to serve this purpose. After 1904, the questions for the promotion examinations—of which military history formed only one of six categories—were selected by the Army Council and generally included one or more of the campaigns described in *Stonewall Jackson*. Every candidate for advancement to any rank up to and including that of major was required to pass these examinations (R. J. Grewing, "Officers," *Encyclopaedia Britannica* [11th ed.; New York, 1910], XX, 20). Perhaps it is significant that one of the members of the Army Council, Major General H. C. O. Plumer, had been a close personal friend of Henderson and was himself convinced of the value of studying the Civil War (see the *Journal of the Royal United Service Institution*, LV [September, 1911], 1144).

26. H. M. E. Brunker, *Story of the Campaign in Eastern Virginia* (London, 1904), was specifically designed to serve as a "crammer" to aid officers in preparing for the examinations. It was little more than a poor condensation of *Stonewall Jackson* and contained numerous errors. J. H. Anderson, *Notes on the Life of Stonewall Jackson* (London, 1905) was written for a like purpose and suffered the same shortcomings. A third book, Edward Nash, *Jackson's Strategy* (London, 1904), is of a higher caliber. Nash showed an independence lacking in most of these campaign studies, and he was among the few who criticized Jackson (pp. 10, 11, 16, 18, 26–27).

27. "Crammers" written to help British officers study for this examination include J. H. Anderson, *Grant's Campaign in Virginia May 1–June 30, 1864* (London, 1908); Charles Francis Atkinson, *Grant's Campaign of 1864 and 1865: The Wilderness and Cold Harbor* (London, 1908); H. M. E. Brunker, *Grant and Lee in Virginia, May and June,*

being tested on their knowledge of the early campaigns in Virginia and Jackson's Valley campaign.[28] In 1912, the questions again were on the Virginia campaigns of 1862 and 1863,[29] while two papers on the Shenandoah Valley campaign—one general and the other covering in detail the events between May 16 and June 9, 1862—were required for the promotion examinations in 1913.[30]

Paradoxically, Henderson, whose primary intent had been to develop a flexible and inquisitive attitude in the minds of his students, ultimately became the agent of a dogmatic approach to the Civil War. Official reliance upon his books and the heavy emphasis upon facts naturally did not encourage much new research or original thought. Most of the campaign studies of this period were little more than abridged—and often mutilated—versions of *Stonewall Jackson*. By writing to drum facts into the heads of candidates for promotion, the authors of these "crammers" succeeded only in distorting the main lessons Henderson had contrived to teach. What had given life to *Stonewall Jackson* was the way in which Henderson had blended his philosophy of war into the narrative. In contrast, the majority of those who followed in his footsteps concentrated on excessive detail and thus ignored the very qualities that had made the

---

1864 (London, 1908); and Captain G. H. Vaughan-Sawyer, *Grant's Campaign in Virginia, 1864* (London, 1908). See also T. Miller Maguire, "The Battle of Spotsylvania," *United Service Magazine*, XXXVIII (October, 1908), 63–71; and two thoughtful articles by Lt. Col. J. E. Edmonds: "The Campaign in Virginia in May and June, 1864," *Journal of the Royal Artillery*, XXXV (1908–9), 521–47; and "Lee and Grant in Northern Virginia in May and June, 1864," *Royal Engineers Journal*, VIII (July, 1908), 5–23.

28. J. H. Anderson, *The American Civil War* (London, 1910), deals only with the Virginia campaigns to June, 1863. See also Major G. W. Redway, "The Shenandoah Valley Campaign," *United Service Magazine*, XL (February, 1910), 522–30 (cited hereafter as Redway, "Shenandoah Valley Campaign"); and Redway, "McClellan's Campaign on the Yorktown Peninsula," *ibid.*, XLI (May, 1910), 166–78.

29. John H. Anderson, *Notes on the Battles of Antietam and Fredericksburg* (London, 1912); Colonel P. H. Dalbiac, *Chancellorsville and Gettysburg* (London, 1911); Colonel J. E. Gough, *Fredericksburg and Chancellorsville* (London, 1913); "Miles" [Walter E. Day], *The Campaign of Gettysburg* (London, [1911]); Captain E. W. Sheppard, *The Campaign in Virginia and Maryland* (London, 1911); H. W. Wynter, "Artillery Fighting at the Battle of Gettysburg," *Journal of the Royal Artillery*, XXXVIII (1911–12), 51–74.

30. T. Miller Maguire, *Jackson's Campaigns in Virginia 1861-2* (London, 1913); Lt. Col. Holmes Wilson, "The Strategy of the Valley Campaign (1861-62)," *Journal of the Royal Artillery*, XL (1913–14), 197–216.

biography of Jackson a valuable military text. A few scattered writers stepped out of Henderson's shadow either by conducting independent research or else by concentrating on campaigns not included in *Stonewall Jackson*,[31] but most of the military studies of the Civil War that appeared in England during this period differed little from the earlier treatment of the Prussian campaigns. As all armies were shortly to discover, "To be able to enumerate the blades of grass in the Shenandoah Valley and the yards marched by Stonewall Jackson's men is not an adequate foundation for leadership in a future war where conditions and armament have radically changed."[32]

Henderson's dramatic portrayal of Jackson together with the sentimentalism of the dying Victorian era did much to produce that legendary atmosphere which in England was beginning to surround both Jackson and Lee. Jackson came to be regarded with an awe and admiration quite out of proportion to his genius. One writer, who had sense enough at least to remain anonymous, openly rejoiced that Stonewall "should be placed upon a pedestal. . . . There may be another and perhaps seamier side to Jackson's career, but Colonel Henderson has done well to keep it from our sight. We want no blemishes or infirmities reproduced upon the statue as it is reared for our delight nor should we tolerate dirt to be flung at it in its glorious completeness."[33]

This volley was aimed at Major C. W. Redway, who had been so bold as to suggest that perhaps too much attention was being paid to the Shenandoah Valley campaign, which was, after all, merely a subsidiary operation and not the decisive campaign of the war. Redway's article was denounced on the ground that it was "bald and dull, paradoxical and misleading,

31. Such works include Captain Cecil Battine, *The Crisis of the Confederacy* (London, 1905), a tactical study of Gettysburg and the Wilderness obviously intended as a sequel to *Stonewall Jackson;* Atkinson, *Grant's Campaigns of 1864;* "Miles," *Campaign of Gettysburg;* and the writings of Edmonds and Redway.

32. Liddell Hart, *The Remaking of Modern Armies* (London, 1927), pp. 170–71. General Fuller writes: ". . . in 1913, I remember a Major recommending Henderson's *Stonewall Jackson* to a brother officer, and then, a few minutes later, when this book was being discussed, committing the error of supposing that [the battle of] 'Cross Keys' was a public house in Odiham and Jackson the name of the man who ran it" (*The Army in My Time,* pp. 53–54).

33. "Hotspur," "Stonewall Jackson: Some Current Criticisms," *United Service Magazine,* XLI (April, 1910), 49.

and subversive to one of the tenets of our military faith." Actually Redway was one of the better English students of the Civil War; he had visited many of the battlefields and was familiar with the *Official Records*. Redway can hardly be blamed for feeling indignant at the reception of his views. "It seems," he wrote in self-defense, "that I had erred in venturing to descant upon a topic which involved reference to . . . Jackson without first consulting the work of . . . Henderson and squaring my views . . . with those of his disciples." There were others who were also critical of Jackson, but these managed somehow to escape the abuse that had caused Redway to wonder, with some justice, if perhaps Stonewall "has not been too fortunate in his biographer."[34]

Redway's case doubtless is an extreme example, but it does illustrate the extent of Henderson's popularity in England. The new interest in the Civil War, which may also have been stimulated by similar experiences of the English in South Africa,[35] the requirements of the Officer Promotion Examinations, and the undue concentration upon the Virginia campaigns all attest to his influence. From Henderson's day until World War I taught them better, British officers concentrated practically all

34. G. W. Redway, "The American Civil War—A Reply," *United Service Magazine*, XLII (March, 1911), 637–39. The article in question was Redway's "Shenandoah Valley Campaign." Evidently others besides "Hotspur" also criticized this article, for Redway wrote: "I have been continually reminded by my critics of the debt we owe to . . . Henderson. One says, 'The Life of Stonewall Jackson is a priceless legacy bequeathed to the British Army': another thinks that 'there is nothing more remarkable than the way in which . . . Henderson's work has been recommended to officers and set over and over again for examinations' " ("The American Civil War," p. 638). Redway's *Fredericksburg: A Study in War* (London, 1906), is one of the better campaign histories of this period.

35. While the Boer War was still in progress, Colonel F. N. Maude had pointed out that the Civil War "deserves far closer and more attentive study" because it approximated "very closely indeed" the conditions of the current struggle in Africa ("Military Training and Modern Weapons," *Contemporary Review*, LXXVII [January–June, 1900], 311). "Miles," who later visited the Civil War battlefields and wrote a respectable book about Gettysburg, thought it "worth considering" that Lord Roberts had promoted Henderson to his staff. To "Miles" this "indicated an opinion on Lord Roberts' part that the particular study of the . . . Civil War was the one that was most immediately applicable to the present Transvaal War" ("Lessons of the War," *Contemporary Review*, LXXVII, 156–57). For others who saw a parallel between the Civil War and the Boer War, see Spenser Wilkinson, *War and Policy* (New York, 1900), chap. xxx; Major E. S. Valentine, "An American Parallel to the Present Campaign," *Fortnightly Review*, LXXIII (August, 1901), 660–67; T. Miller Maguire, "A Study in Devastation," *National Review*, XXXVII (August, 1901), 901–12.

their attention upon one narrow aspect of the Civil War—the campaigns in Virginia—to the general neglect of the campaigns in the west under Grant and Sherman. Those who did trouble themselves to investigate the war in the west concluded that a "disproportionate attention" was being given the eastern theater. They advocated instead a reappraisal of the west, where "the decisive blows were struck" and where the operations were "the most important in a military sense."[36]

These few men were to be vindicated only by a new turn of history, a great war that swept away all the old tactical questions that had cluttered the military literature of the pre-1914 era. Attention was now concentrated on war as a whole, on fundamental questions of strategy and politics, and thus on the neglected aspects of the Civil War. The new generation of British officers who studied the Civil War turned to the campaigns of 1864–65 and the war in the west for instruction. They studied Grant and Sherman as earlier classes had concentrated upon the great Confederate leaders in the east—Lee, and Henderson's Jackson.

Henderson's works survived the war. Unlike the Civil War "crammers," which no longer have a practical value, his books are still read today. Military readers continue to find them useful because the whole approach to the study of military history has changed since 1918. No longer is it fashionable to concentrate on one or two battles and memorize the detailed maneuvers of both sides. The emphasis now is upon problems of strategy and the psychology of generalship, factors which have not been outmoded by recent developments in warfare. Because Henderson looked upon military history as something to stimulate independent thought rather than to provide specific patterns or lessons, the value of *Stonewall Jackson* as a military text has not diminished. Strategy and the psychology of leadership and morale rarely, if ever, change, which explains of course why many English soldiers in the 1930's still thought it worthwhile to visit the Civil War battlefields, particularly the Shenandoah Valley and Chancellorsville, where, book in hand,

36. See W. Birkbeck Wood and Major J. E. Edmonds, *A History of the Civil War in the United States* (New York, 1905), p. 527; John Formby, *The American Civil War* (New York, 1910), I, 72.

they could mentally reconstruct Jackson's most exciting and successful maneuvers.[37]

But Henderson's books are read today for another and equally valid reason: they are good history. *The Campaign of Fredericksburg* is still regarded as a competent study of that battle, while half a century of scholarship and intensified interest in the Civil War has failed to produce a biography of Jackson that replaces Henderson's. Although he was not a professional historian, Henderson certainly was no amateur; his works bear evidence to his early historical training at Oxford. He may have painted Jackson in overbright colors, but these faded when the events of 1914–18 placed the Civil War—and Jackson—in a different perspective. In the main, Henderson's errors were not factual ones. He respected facts, even to the extent that the completed portions of his history of the South African War were ordered scrapped because the facts in this case might have embarrassed the government.[38]

Henderson excelled as a military writer, as a historian, and as a biographer. The student of the Civil War will always regret that he was not spared long enough to write his projected sequel to *Stonewall Jackson*—the life of the greatest of all Southern generals, Robert E. Lee.[39]

37. For an entertaining description of one of these tours, see Lt. Col. A. H. Burne, "How To Visit the Battlefields of America on the Cheap," *Fighting Forces*, XV (April, 1938), 52–60. In his autobiography, John Masters, author of a series of historical novels about India, and himself a former officer in the 4th Prince of Wales Own Gurkha Rifles, writes that in 1938 he spent much time studying Jackson's Valley campaign for the "approaching promotion examination." When he came to the United States the following year, he made a point of visiting the Shenandoah Valley. "It snowed hard, and I took Henderson's *Stonewall Jackson* out of the trunk and lay down on my stomach with it in the blizzard at Port Republic, trying to find out where Tyler had put his guns, and where was the forested path along which the Louisiana regiments had marched to turn the tide of battle" (*Bugles and a Tiger: A Volume of Autobiography* [New York, 1956], pp. 240–41, 274).

38. Maurice, *Sir Frederick Maurice*, pp. 121–24; Robertson, *From Private to Field Marshal*, pp. 118–20.

39. This work was already past the planning stage at the time of Henderson's death. For several years, members of the Lee family had been collecting materials for him to use in his projected biography (Holden, "Lt. Col. G. F. R. Henderson, C.B., In Memoriam," *Journal of the Royal United Service Institution*, XLVII [April, 1903], p. 381; *Southern Historical Society Papers*, II, N.S. [September, 1915], 306). Had Henderson lived to complete this work, it would have been interesting to read his comments on the value of intrenchments. The experiences of the Russo-Japanese War (1904–5) would undoubtedly have intensified his interest in this problem, just as the Boer War had altered his estimation of mounted infantry. Henderson might well have anticipated Major General Sir Frederick Maurice's conclusion that Lee's handling of intrenchments in the campaign of 1864 "should, if it had been studied and appreciated at its true value, have marked an epoch in the history of tactics" (*Robert E. Lee the Soldier*, p. 291).

INDEX

Abolitionists, 176, 177
Adaye, Maj.-Gen. Sir John, 3 n.
Alabama, 194, 221
Aldershot, 210 n.
Aldershot Military Society, 174, 225 n., 254 n.
Aldie Gap, 295
Alexander, Brig.-Gen. E. Porter, C.S.A., 25, 89, 93
Alexander the Great, 119
Allabach, Col. Peter H., U.S.A., 89
Allenby, Field Marshal Lord, 128
Amelia Court House, 168–69
Anderson, Lieut.-Gen. Richard H., C.S.A., 19 n., 25, 54, 64, 97
Anglo-French Entente (1904), 307
Antietam Creek, 14
Appomattox, 197
Aquia Creek, 16, 17, 18, 19, 37, 38 n., 59, 116
Arabi Pasha, 259 n.
Archer, Brig.-Gen. J. J., C.S.A., 26, 61, 62, 78–83, 103, 110
Armistead, Brig.-Gen. Lewis A., C.S.A., 162–63
Army of Northern Virginia, 14, 16, 21, 25–26, 30–36, 37, 42, 61 n., 100, 141, 225, 226, 261, 262, 263
Army of the Potomac, 14, 24–25, 26–30, 31–36, 41 n., 68–69, 96, 100, 104 n., 116, 138, 140, 141, 156, 157, 166, 185, 200, 261, 262, 263, 266, 296

Ashby, Brig.-Gen. Turner, C.S.A., 298, 299
Aston, Maj.-Gen. Sir George, 307–8
Atkinson, Col. E. N., C.S.A., 26, 62, 81–83, 107
Atlanta, 3 n., 126, 127, 195, 201
Austria, 278
  army of, 272, 294, 298

Baker, Gen. Valentine, 104
Baltimore, 233
Banks, Maj.-Gen. Nathaniel P., U.S.A., 18–19, 22, 291, 292, 295, 299
Banks' Ford, 20, 116
Barksdale, Brig.-Gen. William., C.S.A., 25, 54, 55, 56, 63
"Battle of Dorking, The," 211 n.
Battle of Spicheren, The, 2, 121
Battles and Leaders of the Civil War, 122, 130 n.
  reviewed, 130–73, 252
Battles and sieges
  Albuera (1811), 75
  Alma (1854), 270
  Antietam (1862), 14, 30, 41, 81, 98, 110, 147, 153, 154, 160, 287
  Austerlitz (1805), 139, 282
  Beresina (1812), 98
  Bladensburg (1814), 201
  Boteler's Ford (1862), 98–99
  Brandy Station (1863), 191, 214–19, 220, 228
  Bull Run
    first battle of (1861), 32, 135–37, 139,

315

Battles and sieges—*Continued*
  145, 148, 149, 161, 178 n., 183, 207, 289–91
  second battle of (1862), 140, 295
Cedar Mountain (1862), 295
Chancellorsville (1863), 38, 96, 97, 104 n., 209, 221, 225, 287, 289, 296, 297, 298, 313
Chattanooga (1863), 208, 209, 251
Chickamauga (1863), 199, 208
Cold Harbor
  first battle of (1862), 138, 139, 140, 164, 297 n.
  second battle of (1864), 85, 92, 156–57, 275, 278, 281
Cross Keys (1862), 292, 293, 211 n.
Essling (1809), 113
Five Forks (1865), 181 n., 212
Fontenoy (1745), 75
Frayser's Farm (1862), 287, 297
Fredericksburg (1862), 9–119, 140, 155, 157, 204, 205, 222
Gaines Mill; *see* Cold Harbor, first battle of
Gainesville (1862), 297
Gettysburg (1863), 68, 96, 114, 155, 159–60, 164, 166–67, 168, 180, 195, 198, 199, 203, 205, 206, 208, 219, 220, 223, 225–53, 297, 298
Gravelotte (1870), 100–104, 115, 150, 153, 167, 180, 205
Inkerman (1854), 10
Jena (1806), 282
Kernstown (1862), 292, 295, 296, 297
Leipzig (1813), 98
McDowell (1862), 292, 293, 297
Malvern Hill (1862), 140, 162 and n., 203
Mars la Tour (1870), 115, 191, 213
Metz (1870), 150
Murfreesboro (1862), 155
Nashville (1864), 201, 206, 212
Orthez (1813), 113
Paris (1870–71), 150
Plevna (1877), 91, 127, 250, 279
Port Republic (1862), 292, 293, 295, 297, 299, 314 n.
St. Privat (1870), 102, 205, 209
Sedan (1870), 235
Seven Pines (1862), 153, 164
Shiloh (1862), 153, 154
Solferino (1859), 154
South Mountain (1862), 164
Spicheren (1870), 149, 167
Spotsylvania (1864), 208, 268–73, 275, 281, 298
Talavera (1809), 107, 153
Tel-el-Kebir (1882), 2
Trebia (218 B.C.), 111–12
Vicksburg (1803), 194, 195, 200, 226 and n.
Vittoria (1813), 139

Waterloo (1815), 10, 75, 92, 95, 139, 180, 257
Wilderness (1864), 127, 128, 180, 203, 205, 223, 254–82
Winchester (1862), 292, 293; (1864), 220
Woerth (1870), 139, 167, 243
Bavaria, 158
Bazaine, Gen. François Achille, 103, 115
Bealeton, 227, 228
Beaumont, Capt. F., R.E., 179 n.
Beauregard, Gen. P. G. T., C.S.A., 34
Berlin, Va., 14
Bermuda, 2
Bernard's cabin, Fredericksburg battlefield, 61, 78 n., 97 n., 98
Beverly Ford, 214, 219
Big Bethel, 146
Birney, Brig.-Gen. David B., U.S.A., 24, 71 n., 77, 78, 81–83, 84, 85 n., 91, 92, 101 n., 114
"Black Douglas," 266 and n.
Blackwood's Edinburgh Magazine, 4 and n.
"Bloody Angle," Spotsylvania battlefield, 208, 272 n.
Blue Ridge Mountains, 14, 16, 19, 230
Boer War; *see* South African War
Boers, 304
Boguslawski, Capt. A. von, 105
Borcke, Maj. Heros von, C.S.A., 157 n.
Brackenbury, Gen. Sir Henry, 123 n., 304
Bradley, Maj.-Gen. Luther P., U.S.A., 212
Bragg, Gen. Braxton, C.S.A., 155, 286
Branch, Brig.-Gen. L. O'Brien, C.S.A., 26, 62
Brandenburg, 158
Bredow, Gen. von, 213
Bristoe Station, 287
Brooks, Maj.-Gen. William T. H., U.S.A., 25, 77, 83, 101 n.
Brussels, 153
Buell, Gen. Don Carlos, U.S.A. 153
Buford, Maj.-Gen. John, U.S.A. 232–33, 236
Bull Run Mountains, 230, 295
Burgos, 113
Burns, Brig.-Gen. William W., U.S.A., 24, 84
Burnside, Maj.-Gen. Ambrose E., U.S.A., 14, 16–22, 30, 34, 36–43, 52–56, 58, 59, 65–66, 69, 71, 72, 76, 87, 92–95,

96 n., 100, 109, 112–13, 116–17, 155, 266
Butterfield, Maj.-Gen. Daniel, U.S.A., 24

Caesar, Gaius Julius, 11, 156
Caldwell, Brig.-Gen. John C., U.S.A. 24, 75
Cambridge, H.R.H., duke of, 126
Campaign of Fredericksburg, The
discussed, 1–8
significance of, 2–5, 121, 125, 128, 133 n., 314
why written, 5
Campaigns of Hannibal, 111–12
Campaigns in Virginia, Maryland, etc., 3 n., 8, 12; *see also* Chesney, Col. Charles Cornwallis
Canada, 28, 181 n., 182 n.
Cardwell Reforms, 4, 35 n.
Carlisle, Pa., 230, 232, 237
Carthage, army of, 111, 112, 119
Cashtown, Pa., 232, 233
Catlett's Station, 18
Cemetery Ridge, Gettysburg battlefield, 238, 240, 241, 243, 244
Centreville, 18, 136
Chambersburg, Pa., 230 n., 232, 237
Chancellorsville, 20, 21, 209, 228
Charleston, 177
Chesney, Col. Charles Cornwallis, 3, 4, 8, 12, 30, 79 n., 92, 98, 126, 182 n., 197, 211 n.
Chesney, Sir George, 211 and n.
Chesney Gold Medal, 211 n.
Chicahominy River, 138, 162
Cleburne, Maj.-Gen. Patrick R., C.S.A., 184, 186
Cobb, Brig.-Gen. Thomas R. R., C.S.A., 25, 63, 73–74, 76, 106
Combined Training, 306
Contemporary Review, 243
Cooke, Brig.-Gen. John R., C.S.A., 25, 63, 76, 86
Cornwallis, Lord Charles, 137
Couch, Maj.-Gen. D. N., U.S.A., 24
Craufurd, Gen. Robert, 153
Crimean War (1854–56), 4, 257, 259, 298
Cromwell, Oliver, 288
Crutchfield, Col. Stapleton, C.S.A., 61 n.
Culpeper, 16, 17, 214, 216, 219, 227, 228, 262
Culp's Hill, Gettysburg battlefield, 238, 243–45

Cumberland Valley, 226, 231
Cunningham Farm, Brandy Station battlefield, 218
Curragh, 210 n.

Dabney, Maj. Robert L., C.S.A., 98–99, 293 and n.
Dante, 288
Danube River, 278
Davis, Jefferson, 99, 136, 142, 160, 184
Deep Run, Fredericksburg battlefield, 43, 46, 53, 54, 57, 63, 69, 77, 85, 97
Democrat party, 176
"Devil's Den," Gettysburg battlefield, 241, 242
Devonshire, Spencer Compton, eighth duke of, 27 n., 119 n.
Doubleday, Maj.-Gen. Abner, U.S.A., 25, 69, 70, 77, 81, 83, 84, 93, 97, 101, 114, 115
Dublin Military Society, 304
Dumfries, action of (December, 1862), 59, 116
Dunottar Castle, 305

Early, Lieut.-Gen. Jubal A., C.S.A., 21, 22, 26, 42, 57, 58, 62, 81, 85, 99, 240
Easter maneuvers, 149, 171, 172
Edinburgh Review, 120, 130 n., 156
Egypt, 259 n., 305
Ely's Ford, 116
Emmetsburg road, Gettysburg battlefield, 240, 241, 246
Evans, Brig.-Gen. Nathan G., C.S.A., 183
Ewell, Lieut.-Gen. Richard S., C.S.A., 162, 227 and n., 228, 230 and n., 231 n., 232–34, 237, 240, 244, 291, 292

Fairfax Court House, 116
Fairfield road, Gettysburg battlefield, 166
Falling Waters, 289
Falmouth, 17, 18, 21, 22, 36, 38, 40, 44, 46, 50, 54, 64, 66, 113, 117, 225, 227
Farnsworth, Brig.-Gen. Elon J., U.S.A., 248 n.
Field, Maj.-Gen. Charles W., C.S.A., 26, 59, 61
Fisher's Gap, 19
Fleetwood Hill, Brandy Station battlefield, 216–19
Fletcher, Lieut-Col. H. C., 124 n., 138, 141, 142, 146, 160, 178, 182 n.
Forbes, Archibald, 207

Index 317

Forrest, Lieut-Gen. Nathan B., C.S.A., 186, 221
Fort Monroe, 137
Fort Sumter, 134, 177, 193
France, army of, 100–105, 107, 144, 154, 161, 171, 249, 272, 298, 307
Franco-Prussian War (1870–71), 2, 3, 4, 10, 33, 67, 100–105, 110, 115, 120, 123, 126, 132, 149, 150, 158, 162, 167, 201–2, 207, 212, 222, 243, 282, 298
Franklin, Maj.-Gen. William B., U.S.A., 25, 53, 56–59, 66–71, 77, 85, 88, 94, 97, 98, 100, 101, 114, 115, 117, 297
Frederick, Md., 230, 233, 236
Frederick II (the Great), 11, 207, 209
Fredericksburg, 9–119 *passim*, 226–28, 267
Freeman, Douglas Southall, 8 n., 43 n., 44 n., 79 n., 84 n., 97 n., 159 n., 228 n., 302 n., 303 n.
Fremantle, Lieut.-Col. James Arthur Lyon, 6 n., 178
Frémont, Maj.-Gen. John C., U.S.A., 292, 295, 299
French, Maj.-Gen. William H., U.S.A., 24, 72–73, 76, 77, 86
From Manassas to Appomattox, 225 n.
Front Royal, Va., 228, 298, 303 n.
Fuller, Maj.-Gen. J. F. C., 3, 306, 311 n.

Gambetta, Leon, 154
Garibaldi, Giuseppe, 28
Georgia, 126, 194, 195, 199, 201, 221
troops of
16th Regiment, 63, 88
18th Regiment, 63
24th Regiment, 63, 77
Phillip's Legion, 63, 77, 88
German Empire, 178
German Legion, at Talavera, 107
Germany, army of, 2, 4–5, 101–5, 121, 123, 148–50, 158, 162, 167–69, 171, 179, 205, 207, 213, 243, 255, 272, 298, 307
Getty, Maj.-Gen. George W., U.S.A., 24, 55, 89
Gibbon, Maj.-Gen. John, U.S.A., 25, 69, 77, 78, 80–85, 92, 101 n., 106, 114, 297
Gordonsville, 16, 17, 51, 287
Grant, Gen. Ulysses S., U.S.A., 29, 34, 42, 85, 92, 96, 127, 128, 148, 155, 156–57, 161, 168, 180, 185, 185–87, 194–96, 199–203, 207, 224, 251, 258, 261–64, 266, 268–69, 271, 273–75, 278–80, 289, 296–98, 309, 313

Great Britain
army of, 7, 122, 179 n., 180, 201, 304, 305, 307
Auxiliary forces, 122, 123
volunteers, 5–7, 10, 119 n., 123–24, 169–73, 188, 211 n.
yeomanry, 126 n., 211 n., 223
Parliament, 171
terrain, 191
Gregg, Brig.-Gen. Maxcy, C.S.A., 26, 61, 62, 78, 79, 103, 110
Griffin, Brig.-Gen. Charles, U.S.A., 24, 86
Guiney's Station, 22, 39, 43, 57

Haig, Field Marshal Sir Douglas, 128
Hale, Col. Lonsdale, 3 and n., 243
Halifax, 2
Halleck, Maj.-Gen. Henry W., U.S.A., 155
Hamilton's Crossing, Fredericksburg battlefield, 50, 54, 57, 59, 62, 64
Hamley, Lieut.-Gen. Sir Edward, 3–4, 102, 178 n., 302–3
Hampton, Lieut.-Gen. Wade, C.S.A., 22, 26, 59, 64 n., 214
Hancock, Maj.-Gen. Winfield Scott, U.S.A., 8, 24, 33, 73–77, 86, 236–37, 258
Hannibal, 111–12, 119
Hanover Court House, 20
Harper's Ferry, 140, 287, 296
Harrisburg, 230, 237
Harrisonburg, 292, 295
Hartington, Marquis of; *see* Devonshire, Spencer Compton, eighth duke of
Harvard, 184 n.
Havelock-Allen, Sir Henry, 181 and n.
Hays, Maj.-Gen. Harry T., C.S.A., 26, 62, 81
Hazel Run, Fredericksburg battlefield, 46, 47, 48, 52, 54, 57, 63, 72, 77, 86, 88, 89, 94, 97, 117
Hazen, Maj.-Gen. William B., U.S.A., 148, 160–61, 203, 207, 209
Heidlersburg, Pa., 233–34
Heilsberg, Russian intrenched camp at (1807), 92
Henderson, Col. George Francis Robert
correspondence cited, 8 n., 121 n., 121–22, 129, 162 n., 184 n., 284, 285, 287 n., 297 n., 302 n., 304 n.
early career, 2
influence of, 305–14
later career, 306
later writings, 122–28

military views
  conscription, 79 n., 172–73
  discipline, 7, 27, 31, 119, 138–60, 169–73, 188, 213–14
  leadership and staff duties, 6, 7, 27–29, 33, 34, 67, 102–5, 114, 119, 159–73, 182–87, 192, 229, 237, 238–40, 247–51, 259–60, 307
  morale, 28, 132–34
  strategy, 39–40, 197–201, 229–30, 293–95, 296, 303–7
  tactics
    artillery, 97, 106, 110–11, 203–6, 251–53
    cavalry, 7, 32–33, 35, 109, 115, 121, 124–26, 179, 190–92, 210–24, 236, 298–99
    raids, 116 and n., 231
    infantry
      defensive, 50, 76, 84 n., 95–96, 99–100, 103, 104–8, 279–83; *see also* fire discipline; intrenchments
      fire discipline, 7, 75–76, 106–9, 159, 189, 209
      intrenchments, 7, 35, 126–28, 265–66, 279–83, 314 n.
      marching, 29, 34, 35 n., 145, 147–51, 189, 299–300
      offensive, 7, 91–92, 103–5, 121, 207–9, 269–70, 271–73
      outpost duties, skirmishing and reconnaissance, 22, 70, 109–10, 151, 164–65, 189, 234
      training, 10, 103, 104–5, 182–83
      volunteer soldier, 119, 123–24, 134–73, 189, 255–59
  reasons for writing *The Campaign of Fredericksburg*, 5–6
  in South Africa, 304–5
  at Staff College, 120
  Stonewall Jackson and the American Civil War, 301–4

Henry Hill, Bull Run battlefield, 290
Hewett, Capt. Edward, British observer, 27 n.
Hildyard, Col. H. T. J., 308
Hill, Lieut.-Gen. Ambrose Powell, C.S.A., 22, 23 n., 26, 37, 39 n., 58, 61, 62, 70, 71, 78–85, 98, 103, 227, 228, 230 and n., 233, 234, 237, 246
Hill, Lieut.-Gen. Daniel Harvey, C.S.A., 21, 22, 26, 42 n., 43, 57, 58, 62, 71, 81, 85, 98, 103, 186
Hohenlohe-Ingelfingen, Prince Kraft zu, German military writer, 150
Hoke, Maj.-Gen. Robert F., C.S.A., 26, 62, 81–83, 107
Home, Col. Robert, 104, 110, 114

Hood, Lieut.-Gen. John B., C.S.A., 19 25, 54, 57, 63, 81, 84, 98, 107, 111 166, 286
Hooker, Maj.-Gen. Joseph, U.S.A., 24, 25, 29, 38, 42, 53, 55, 56, 66, 77, 86, 88, 89, 95, 106–7, 109, 115, 116, 153, 157, 222, 226–30, 232, 235–36, 266, 287, 296, 298
Hop Yard, 22, 37, 38, 42, 43, 58
Hotchkiss, Maj. Jedediah, 8 n., 121, 162 n., 184 n., 284–85, 287 n., 297 n., 301, 302 n., 304 n.
Howard, Maj.-Gen. O. O., U.S.A., 24, 55 n., 77, 86, 92
Howe, Brig.-Gen. Albion P., U.S.A., 25, 77, 83, 101 n.
Howison's mill, Fredericksburg battlefield, 63
Humphreys, Maj.-Gen. Andrew A., U.S.A., 24, 33, 86, 88, 90
Hunt, Maj.-Gen. Henry J., U.S.A., 25, 29, 252–53
Hutton, Maj.-Gen. Sir Edward, 210 n.

India, 4, 11, 181, 182, 308, 314 n.
Indian Mutiny (1857), 4
Infantry Drill Book, 306
Irish Brigade, 24, 73–76; *see also* Meagher, Brig.-Gen. T. F.
Italian Campaign (1896), 294
Italy, army of, 272

Jackson, Mary Anna (Mrs. T. J.), 284, 296, 297, 304
Jackson, Lieut.-Gen. Thomas Jonathan ("Stonewall"), C.S.A., 8, 16, 19–21, 33, 34, 54, 59–63, 68, 72, 83, 93, 96, 98–99, 102, 104, 115, 118, 127, 140–41, 147, 150, 155, 162, 164–65, 183, 185–86, 198, 200, 209, 251, 258, 288–300, 301–3, 305, 310–12
James River, 17, 138, 162, 278, 281, 292
Johnson, Maj.-Gen. Edward, C.S.A., 240, 243–45
Johnston, Gen. Joseph E., C.S.A., 34, 164, 185, 286, 291–92
Joinville, François Ferdinand, Prince ed, 144 n.
Jomini, Baron de, 101 n.
Jones, Brig.-Gen. John R., C.S.A., 26, 62
Jones, Brig.-Gen. William E., C.S.A., 214
Junot, Gen. Andoche, 157

Kelly's Ford, 116, 214, 219
Kemper, Maj.-Gen. James L., C.S.A., 90

Kenly, Bvt. Maj.-Gen. John R., U.S.A. 292, 298
Kennedy, Gen. Sir James Shaw, 10–11
Kensington, 3
Kentucky, 193, 194, 201
Kershaw, Maj.-Gen. J. B., C.S.A., 25, 63, 76–77, 88, 106
Kinglake, Alexander William, English military historian, 80
Kossuth, Lajos, Hungarian patriot, 28

La Haye Sainte, Waterloo battlefield, 153
Lane, Brig.-Gen. James H., C.S.A., 26, 61–62, 78–83, 103
Law, Brig.-Gen. Evander M., C.S.A., 241
Lecomte, Lieut.-Col. Ferdinand, Swiss military observer, 27 n.
Lee, Maj.-Gen. Fitzhugh, C.S.A., 22, 26
Lee, Gen. Robert E., C.S.A., 6–8, 14, 16–17, 19–21, 30–33, 36–40, 42–43, 48–49, 52–53, 57–59, 61–65, 67–68, 72, 81, 84, 88, 90, 93–100, 104, 105–11, 112, 114, 116, 118–19, 127, 137, 140–41, 153, 155, 159 n., 160, 162, 166, 168, 181 n., 182 n., 185–86, 194, 196–200, 204–5, 207–9, 222, 224–26, 228 and n., 229–31, 231 n., 232, 234–36, 238, 242, 245–50, 250 n., 251–52, 258, 261–69, 271, 273–75, 278–80, 282, 285–87, 289, 291–96, 298, 300, 311, 314
Lee, Maj.-Gen. W. H. F., C.S.A., 22, 26
Lee's Hill, Fredericksburg battlefield, 44, 46–49, 54, 63–65, 88, 110
Leesburg, 230–32
Leetown, 230
Lexington, 34
Liddell Hart, Capt. B. H., 303
Light Division (Civil War), 26, 58, 61, 79 n., 85; *see also* Hill, Lieut.-Gen. Ambrose Powell,
Light Division (Peninsular War), 144, 153
Lincoln, Abraham, 14, 17, 38, 100, 116, 129, 134, 137, 156–57, 176–77, 185, 193, 195, 202, 226, 229–30, 292, 296, 304
Lisbon peninsula, 280 n.
Little Round Top, Gettysburg battlefield, 238, 240–48
Loire River, 154
London, England, plans for defense of, 123 n.
Longstreet, Lieut.-Gen. James, C.S.A., 16–22, 25, 39, 48, 54, 57, 59, 65, 72, 88, 90, 97, 106, 115, 164–66, 186, 199, 208, 225 n., 227 and n., 228, 230 n., 234–35, 238 and n., 240–42, 242 n., 244, 246–48, 250 n.

Louisiana, 198
troops of, Washington Artillery, 80

McClellan, Gen. George B., U.S.A., 14, 16, 27, 28, 34, 137–38, 140, 146, 153, 155, 161–62, 179 n., 186, 287, 291, 295
MacDougall, Gen. Sir Patrick, 3, 111–12
McDowell, Maj.-Gen. Irvin, U.S.A., 42, 136, 148–49, 161, 292, 295
Macedonian phalanx, 119
MacKenzie, Brig.-Gen. Ronald S., U.S.A., 186
McLaws, Maj.-Gen. Lafayette, C.S.A., 19 n., 25, 54, 63, 166
MacMahon, Marshal, 150, 235
McMillan, Col. Robert, C.S.A., 73 n., 75
Madison Court House, 19, 20
Mago, 112
Mahan, Capt. A. T., 211 n., 305
Maine, 141
Maoris, 181 n.
Marye's Hill, Fredericksburg battlefield, 8, 44, 46–48, 50, 52, 54, 63–66, 73–77, 85–86, 88–91, 93–94, 102–3, 106, 117
Marye's House, 60
Maryland, 2, 14, 140, 193–94, 222, 287
Massachusetts troops
  19th Regiment, 55 n.
  20th Regiment, 55 n.
  28th Regiment, 73 n.
  29th Regiment, 73 n.
Massanutton Mountains, 292
Massaponax River, 44, 46, 49, 57, 58, 64–65, 97–98, 106
Masters, John, 314 n.
Mattaponi River, 36
Maude, Col. F. N., 5 n.
Maurice, Maj.-Gen. Sir Frederick, 96 n., 228 n., 314
Maurice, Maj.-Gen. Sir John Frederick, 1, 2, 4–5, 130 n., 133, 158 n., 159 n.
Maurin, Capt. Victor, C.S.A., 63, 73
Meade, Maj.-Gen. George G., U.S.A., 25, 33, 42, 69–72, 77–86, 92, 101 n., 102–3, 107, 114, 195, 198, 230, 234–36, 238, 240, 247–48
Meagher, Brig.-Gen. Thomas F., U.S.A., 24, 73, 74 n.
Mechanicstown, Pa., 233
Mechum River Station, 292
Mediterranean, 256
Mexican War (1847), 33, 184, 204
Michie, Brig.-Gen. Peter S., U.S.A., 221
Michigan troops, 7th Regiment, 55 n.
Middletown, 303 n.

Milford, 16
Military Magazine, 120
Miller; *see* McMillan, Col. Robert
Miller House, Brandy Station battlefield, 217
Mine road, Fredericksburg battlefield, 69, 85
Mine Run, 262, 265
Mississippi River, 193-94, 198-200, 226, 261
Mississippi Valley, 190, 194
Missouri, 193-94
Modern Warfare as Influenced by Modern Artillery, 3
Moltke, Count Helmuth von, 3, 148, 167, 177, 178 n., 200-201, 260, 305
Moselle River, 102
Mountain Run, Brandy Station battlefield, 219
"Mud March," 116 and n.

Napier, Lieut.-Gen. Sir William, 107
Napoleon I, 6, 11, 67, 92, 95, 112-13, 133, 146, 153, 158, 167, 182, 197, 202, 207, 229, 245, 254-55, 257, 259, 280, 282, 293-94, 296-98, 300.
Napoleon III, 154
New Orleans, 194
New York troops
  63d Regiment, 73 n.
  69th Regiment, 73 n.
  88th Regiment, 73 n.
  89th Regiment, 55 n.
New Zealand, 181
Newton, Maj.-Gen. John, U.S.A., 25, 77, 81, 84, 101 n.
North Anna River, 20, 36, 99, 273-74, 281
North Carolina, 195
  troops of
    7th Regiment, 61
    24th Regiment, 63, 77
Notes on Waterloo, 10-11
Ny River, 268

Oak Shade Church, Brandy Station battlefield, 218
Officer Promotion Examination, 309-10, 312
Official Records of the Union and Confederate Armies, 8, 131, 301, 312
Ohio, 141
  troops of, 41st Regiment, 160
Old Stage road, Fredericksburg battlefield, 47, 50, 53, 57, 64, 66, 68-70, 78, 80, 82, 93-94, 98, 100

Operations of War Explained and Illustrated, The, 4, 178 n., 302-3
Orange and Alexandria Railroad, 14, 16, 20, 263
Orange Court House, 20, 21, 262, 264
Orange plank road, 47, 48, 54, 63, 72, 90
Ordinance of Secession, 32, 176-77
Osman Pasha, 250
Oxford University, 2, 308, 314

Palfrey, Brig.-Gen. Francis W., U.S.A., 170
Paris, Louis Philippe, Comte de, 131 n., 138, 144, 157, 178, 234
Paxton, Brig.-Gen. Elisha F., C.S.A., 26, 62, 81
Peach Orchard, Gettysburg battlefield, 238-41, 243, 245
Pelham, Maj. John, C.S.A., 70-71, 78, 106
Pender, Maj.-Gen. W. Dorsey, C.S.A., 26, 61, 80
Pendleton, Col. Edmund, C.S.A., 62 n.
Pendleton, Brig.-Gen. William N., C.S.A., 26, 48, 187
Peninsular campaign (1862), 144 n., 145, 179 n., 222, 291
Peninsular War, 257, 259, 299
Pennsylvania, 141, 195, 198, 226, 231 n.
  troops of, 69, 103, 105
    116th Regiment, 73 n.
Petersburg, 168, 196, 279
Pettigrew, Brig.-Gen. Johnston J., C.S.A., 246
Philadelphia, 233
Philip's House, Fredericksburg battlefield, 52
Phillip's Legion, 63, 77, 88
Pickett, Maj.-Gen. George E., C.S.A., 19, 25, 54, 57, 63, 81, 84, 90, 98, 111, 246-48, 253
Pipe Creek, 237
Plato, 156
Pleasonton, Maj.-Gen. Alfred, U.S.A., 214
Plumer, Maj.-Gen. H. C. O., 309 n.
Po River, 268
Pollock's Mill, Fredericksburg battlefield, 54
Pope, Maj.-Gen. John, U.S.A., 42, 140, 150, 155, 164-65, 222, 286-87, 295
Port Royal, Va., 21-22, 36-38, 42 n., 43, 47, 58, 65-66, 278
Porter, Maj.-Gen. Fitz-John, U.S.A., 138
Portugal, 157, 297

Index 321

Potomac River, 14, 16–18, 36, 59, 116, 193–94, 202, 225–26, 230, 232–33, 247, 249, 267, 290, 292, 299
Prospect Hill, Fredericksburg battlefield, 44, 59, 62, 93, 97 n., 98
Puritans, 176

Quatre Bras, 180

Ransom, Maj.-Gen. Robert, C.S.A., 25, 54, 63, 75, 86
Rapidan River, 17, 36, 38, 262–64, 279, 281
Rappahannock River, 16–18, 20–22, 36–38, 40, 42, 44, 46–47, 52, 54–56, 84, 94, 97–99, 110–11, 115–17, 204, 214, 218–19, 225–29, 267, 292
Redway, Maj. C. W., 311–12
Republican party, 176
Reynolds, Maj.-Gen. John F., U.S.A., 25, 114
Rhett, Capt. A.B., C.S.A., 64
Richmond, 14, 16–17, 22, 31, 36, 40–42, 47, 50–51, 110, 112, 137, 140, 168–69, 193–94, 196–97, 200–201, 203, 225, 227–28, 261–64, 267, 274, 278–80, 286, 289, 292–93
Richmond and Potomac Railroad, 16, 19, 22, 37, 47, 50, 54, 71, 80
Roberts, Field Marshal Lord, 304–5, 306, 312 n.
Robertson, Field Marshal Sir William, 128
Rodes, Maj.-Gen. Robert E., C.S.A., 240
Roman army, 111–12
Romney, 291
Ross, Gen. Robert, 201
Rosser, Maj.-Gen. T. R., C.S.A., 22, 26, 64 n.
Royal United Service Institution, 303
Russia, 178
    army of, 91, 102, 104, 270, 298
Russo-Japanese War (1904–5), 314 n.
Russo-Turkish War (1877–78), 127
Russy, Capt. Gustavus A. de, U.S.A., 66, 101

Sailor's Creek, 212
St. James' Church, Brandy Station battlefield, 214, 216, 218
Sandhurst, 2, 120, 202
Savannah, 195, 201
Schaw, Col., H., 99
Scheibert, Major Justus, Prussian military observer, 6 n., 108 n., 157 n., 220 n.
Science of War, The, 122, 130 n., 306

Semmes, Brig.-Gen. Paul, C.S.A., 25, 63
Seneca, Lucius Annaeus, 112
Seven Days' Battles (1862), 137, 140, 146, 287, 297 n.
Seven Weeks' War (1866), 121, 132, 162, 167, 179
Sharpsburg, battle of; *see* Battles and sieges, Antietam
Shenandoah Valley, 8, 14, 18, 19, 147, 150, 193, 220, 224, 226, 228, 232, 249, 286–87, 291–92, 302, 309, 313
Shenandoah Valley Pike, 303 n.
Shepherdstown, 230
Sheridan, Maj.-Gen. Philip H., U.S.A., 33, 125, 181 n., 185–86, 200, 211–12, 220–21, 224, 258
Sherman, Gen. William T., U.S.A., 3 n., 126–27, 148, 178 n., 186–87, 195, 199, 201, 208–9, 224, 255, 258, 313
Sickles, Maj.-Gen. Daniel E., U.S.A., 24, 71 n., 83–85, 101 n.
Sigel, Maj.-Gen. Franz, U.S.A., 18, 22
Skinker's Neck, 21–22, 37–39, 43, 53 n., 59
Skobeleff, Gen. Mikhail Dmitrievich, 67, 207
"Slaughter-pen," Fredericksburg battlefield, 75, 90
Smith, Maj.-Gen. William F., U.S.A., 25
Smithfield, Fredericksburg battlefield, 97
South African War (1899–1902), 7, 79 n., 126–27, 304–5, 312 and n., 314 and n.
South Anna River, 36
South Carolina, 176–77
    troops of
        3d Regiment, 106
        15th Regiment, 88
        Orr's Regiment of Rifles, 79 n.
South Mountain, 226, 234, 235
Spain, 299
Spotsylvania County, 22, 43
Spotsylvania Court House, 19
Staff College, 1–3, 120, 122, 307–8
Stafford County, 37, 44
Stafford Heights, Fredericksburg battlefield, 43, 46, 48, 50–54, 59, 66, 68, 71–72, 93, 96 n., 97, 100, 101, 115, 117
Stanhope Memorandum, 123 n.
Stansbury's Hill, Fredericksburg battlefield, 44, 46, 48, 64, 73
Starke, Brig.-Gen. William E., C.S.A., 26, 62 n.
Stevensburg, 219
Stone Bridge, Bull Run battlefield, 290
Stoneman, Maj.-Gen. George, U.S.A., 24
Stonewall Brigade, 290

Stonewall Jackson and the American Civil War, 120, 301–4, 306, 308, 310–11, 313

Stuart, Maj.-Gen. James E. B. ("Jeb"), C.S.A., 17–18, 21–22, 26, 33, 38, 50, 57, 59, 61, 64, 70–71, 77, 81, 84, 93, 98, 109, 115–16, 125, 186, 211–13
at Brandy Station, 214–19, 220–24, 228, 230, 231 and n., 232, 248–49, 250, 298

Sturgis, Maj.-Gen. Samuel D., U.S.A., 24, 86, 88

Sumner, Maj.-Gen. E. V., U.S.A., 18, 24–25, 53, 55 n., 56–58, 66–67, 72, 77, 84, 88, 106–7, 113, 115, 117

Susquehanna River, 230–31

Swinton, William, 288

Sykes, Maj.-Gen. George, U.S.A., 24, 86, 89

Taliaferro, Brig.-Gen. William B., C.S.A., 62, 81, 85

Taylor, Lieut.-Gen. Richard, C.S.A., 147, 184 n.

Taylor's Hill, Fredericksburg battlefield, 44, 54, 64, 73

Telegraph road, Fredericksburg battlefield, 47–48, 50, 63–64, 72, 77, 89–90

Tennessee, 194, 201

Terry, Maj.-Gen. Alfred H., U.S.A., 186

Texas, 193, 198

Thomas, Brig.-Gen. E. L., C.S.A., 26, 61–62, 81–82

Thompson's House, Brandy Station battlefield, 218

Times, the, 75, 120, 305, 306, 308

Torres Vedras, lines of, 113, 279–80, 280 n.

"Trognon, A"; *see* Joinville, François Ferdinand, Prince de

"True Reformer, A"; *see* Maude, Col. F. N.

Turkey, army of, 127

Tyler, Brig.-Gen. Erastus B., U.S.A., 89

United States Army, Sixth United States Cavalry, 216

United States Mine Ford, 20–21, 116

Versailles, 148

Vienna, Va., 136

Virginia, 2, 3, 14, 36, 117–18, 135, 140, 180, 190–91, 193–95, 198, 200–203, 214, 226, 229–30, 247, 259, 262–63, 267, 287, 291–92, 299, 303, 309–10
troops
9th Virginia Cavalry, 218
10th Virginia Cavalry, 218
12th Virginia Cavalry, 216
13th Virginia Cavalry, 218

Virginia Central Railroad, 51

Virginia Military Institute, 186

Walker, Brig.-Gen. James A., C.S.A., 26, 62, 81–82

Walker, Brig.-Gen. R. Lindsay, C.S.A., 26, 59, 61, 71, 80, 93

Walton, Col. James B., C.S.A., 25, 63, 73–75, 89, 93, 117

Wardensburg, Va., 299

Warren, Col. E. T. H., C.S.A., 26, 62

Warren, Maj.-Gen. Gouverneur K., U.S.A., 240, 243

Warrenton, 14, 16, 18

Washington, D.C., 16–18, 22, 38, 116, 131 n., 135, 137, 140, 193, 198, 201, 226, 228–30, 233, 235, 261, 267, 278, 289–92

Waterloo Lectures, 3 n.

Wellington, Arthur Wellesley, first duke of, 10–11, 42, 67, 92, 107, 144, 255, 260, 280 n., 288, 295, 300, 307

West Point, United States Military Academy at, 34, 134, 182 n., 184 and n., 186

West Virginia, 231 n., 291 n., 292

Whipple, Brig.-Gen. Amiel W., U.S.A., 24, 66, 72, 77, 86

White House on Pamunky River, 278

White Oak Swamp, 297

Wilcox, Maj.-Gen. Cadmus M., C.S.A., 19 n., 246

Willcox, Maj.-Gen. Orlando B., U.S.A., 24

Williamsburg, 163–64

Williamsport, 230

Willoughby Run, Gettysburg battlefield, 236

Wilson, Field Marshal Sir Henry, 128

Wilson, Maj.-Gen. James H., U.S.A., 221

Wisconsin, 141, 142

Wolseley, Field Marshal Lord, 1, 2, 4–5, 108 n., 123, 125, 126 n., 130 n., 160, 178, 185, 189, 191, 210 n., 285, 304, 306–7

Yale, 184 n.

Yerby's House, Fredericksburg battlefield, 22, 57, 64

York, Pa., 230, 232–33

York and Lancaster Regiment, 2

York River, 138, 140

Yorktown, 137, 146

Zook, Col. Samuel K., U.S.A., 24, 73, 75